UFOs and the Nature of Reality

Understanding Alien Consciousness and Interdimensional Mind

Edited by
Judi Pope Koteen

Published by
Indelible Ink
P.O. Box 1415
Eastsound, WA 98245
Phone (206)376-5184

Cover and interior design by Debbi Burke Edkins
Typesetting by Daniel Ferris, Athens Enterprises
Patience, understanding, and immense support by Carol Scarfuto and Minnie Tuck
Thanks to Marri Parkinson, Carol Wright, Tom Kelty, and Beth Weber
Printed in the United States of America
by Gilliland

Library of Congress Catalog Card Number: 90-083741
ISBN: 0-9627267-4-5

If you think, out of ten billion suns in the
Milky Way, that this pitiful speck of dust, this
emerald with the yellow sun
in the outermost corner
of God's mind, is the only planet to host life,
you need a vision!

Dedication and Thanks

This book is dedicated to my best friends, my children, Jennifer and Adrianne, who have endured what to many would seem like immense personal sacrifices to support my dream. A dream which has taken them beyond the bounds of earthly thinking and normal acquisitions, out of the city to an island in the Northern Sky. They have traded lunch at the country club for pot lucks at the Grange. Their ballet lessons are now given by the Great Blue Heron as she gracefully moves across the sky, despite her awesome appendages. They are learning to soar chasing the eagle's shadow along the beach.

They trusted me to take them to someplace else, and I will always love them.

They are children of light, tolerance, and great beauty.

Foreword

In 1977, Ramtha appeared before JZ Knight in the doorway of her kitchen in Tacoma, Washington. Nothing has been the same since.

Ramtha is an enigma. He calls himself such. And without a doubt, as the focus of society moves beyond the phenomenon of channeling and more people begin to take a serious look at the message, Ramtha will be hailed as one of the greatest communicators and teachers of all time.

He lived on earth 35,000 years ago and for the better part of that life was a savage conqueror, until he was wounded in a battle. He withdrew to a mountain to contemplate life and spent seven years upon a barren rock, considering the sun and moon, and life and death, and things that never seem to change. And when he left that rock and returned to his waiting army, he returned an enlightened being, for he had seen through the illusion called life. He stayed and taught his people for many years and ultimately ascended before thousands. He is "The Ram" around which the Hindu people built their religion.

He describes himself as part of a brotherhood of beings who love humankind greatly, who have come with information to help us make the choices necessary in order for us to gain personal sovereignty in the days to come. His is an enabling message, an empowering message.

Once, when asked what he would say if he could reach the whole world with a single most important message, he said, "I would tell you that you are God and that you are greatly loved."

Books created from Ramtha's words are based on events called Intensives and Retreats, where he speaks extemporaneously. This book is based on one Intensive and one Retreat specifically addressing the subject of UFOs and interdimensional understanding.

Editor's Notes

A "voice" came to ancient people and we still worship that information because it's what we've always been told. Major religions and the fate of men and empires have been determined by interpretation of ancient scripture based on that voice.

How do you think that "voice" came to those people? If a "voice" came to you today, would you think it was God? Would you listen? Why would a voice that came to you today be less important than a voice that came to people you never met over 2,000 years ago? I am not trying to discredit ancient information. I am simply trying to encourage you to put it in perspective. Everything is relative.

I want you to read this book because I am concerned about where societal consciousness is today. I am concerned about a closed-minded mentality that attempts to prevail through tactics of fear and dogma. And because I have always believed that if you deliver the truth to people, they will recognize it, and they will listen.

If you have read a lot of similar material, this book will broaden your understanding of alien consciousness and tie together a lot of material you've read, answering many questions you've held. If this is the first book on this subject you've ever picked up, I suggest it only be a beginning. Because for you, it will raise as many questions as it answers, purposefully. And it will start you on a magical journey that will carry you to the headwaters of truth and a whole new understanding of creation and God.

The material in Section One is based on an Intensive called UFO: Raiders From Above. The material in Section Two was delivered in Yucca Valley and was titled UFO Interdimensional Understanding: Making Contact.

In the final chapter, references to Consciousness and Energy refer to a specific process Ramtha teaches. That process cannot be communicated in a book. There are no words. But the understanding that thought creates reality by virtue of applied consciousness and applied energy is critical to the comprehension of much of this material, therefore I have included some of these references but made no attempt to describe the process.

Table of Contents

SECTION TWO

UFO Interdimensional Understanding:

Making Contact Yucca Valley

First and foremost, you should know what you are. You should know that before you ever start looking around to discover anyone else.

You are made up of a cosmology that defies planetary explanation because it is the essence of planetary explanation. It is called God. It is consciousness and energy. The life force.

Your intelligence and your capacity for intelligence is equal to anything that lies in any star system, in any dimension. You have the keys to be it all. Knowing yourself first, before you go around looking under rocks and bushes and brambles is very, very important.

These entities exist. Their greatness is legend. What they are capable of doing is awesome. And what they can share with you is a word called magnificent.

Section One

Raiders From Above

Introduction to Raiders From Above

You picked up this book because you perceived it would enhance your understanding of alien consciousness. Not "this world" social consciousness, but alien consciousness. You hoped to understand what the "raiders" are all about. And secretly, some of you wondered if perhaps you had already been induced by them.

Let's address that right away. There are a lot of people reading this who have already been raided.

There are many things I can't tell you here because you are not ready to hear it. Some of you are not very grown up, not very mature in consciousness, and as a result, you always ride the brink of hysteria. You are prone to fanaticism. You shift on a sheckel very quickly. You are highly strung and highly emotional. So for those of you reading this who fit that category, you will simply not be able to know all there is to know about these people, their strain, and their ways of consciousness, which are, in essence, truly alienated from your own.

Then there are those of you who don't ride the brink of hysteria. You, for the most part, are seekers for the stars in the sky. You present a communication difficulty as well, because you create a whole dogma around the concept of alien contact.

Here's how that works. You think if they contact you, that makes you special. You rather worship them. You think where they go must be much nicer than where you are. You want them to rescue you from your dilemma, which of course, you, yourself, have created. (Your reality is created by you, not someone else!) As a result of your tendencies to worship, I must be mindful of what I tell you regarding the ways of some of the aliens, because you would make it your own religion, and that is not why this information is being put before you.

And then last but far from least, there are a few of you who are very astute. You are becoming very centered in your own power. You are finding your own consciousness. You are finding your own deliverance. You have a centering. You are not hysterical, nor are you prone to dogmatic truth. You will not make these aliens, your brothers, the hierarchy of your life. Many of you are the ones who have had contact, simply because of the stability of who you are.

For you see, these entities, these aliens as you call them, are not interested in hysterical followers. They are interested in ingenious human beings who have the ability for knowledge, beings who have great interior strength, beings who have a mystical understanding of spiritual nature, beings who exhibit a natural quest for scientific understanding and a basic desire to know, beings to whom *knowing* is an integral part of destiny. These are the kind of people the aliens you would *want* to have contact with, are interested in contacting.

And there is another group of human beings that you might belong to. This is a group, it is sad to report, that has conspired, throughout history, to create dogma and bitterness. But their plan will go awry, like most human beings' plans in the long run of tyranny.

It is with some reluctance that I deliver this information about my brothers. And since I am addressing representatives from all of the above groups, I will deliver this information with tact, bearing in mind the blending of energies within the readership. I desire to give you a favorable understanding, to give you knowledge, so that you are not running around in fear and ignorance. And for some of you, who can really know the truth, your contact has already been arranged, because you have made yourselves worthy of it, and because you have an attitude that is not a pretense, but is based solidly on the character of who you are.

So, I allow these words to be written with some reluctance because this book will not be read by a cohesive group, but rather by an assortment of people coming from a vast array of understandings. With that in mind, I want you to understand that you will not be told everything because you are not ready for everything. And no, all of you are not going to have the experience of contact because you are not worthy of it. Mainly because you are not worthy of yourself and you have not made the path right within you.

But I will give you a lot of information which, when coupled with the knowledge just imparted, will allow you to gather an understanding. Then one day when you look up into the skies you'll be able to see what's there without trepidation, without worship, and without needing to make yourself feel special. The doors of contact will open to you when your attitude is solidly based on a simple desire to share.

This knowledge you are about to acquire will be very valuable in your future, because this is not knowledge coming from an observer at the human level, but is, instead, coming from one at another level. I know who these people are. I know why they are here. I know why they've been here. And they know who I am. I work in harmony with them.

These entities are very different from you in personality and emotional structure. They are very different from you in the physical sense of the word. There are those of you who would find my people very repugnant because of their coloration, skin tissue, and structural build.

Like I've always said, beauty is truly an unseen essence. It is nothing of the flesh. The flesh is merely a genetic purpose to endure environment.

The people I will principally tell you about here are the Great Workers.

There are many strains and many brotherhoods, just as there is certainly an infinite galaxy of worlds. Your Milky Way, as you call it, has ten billion suns in it. And how many of those are universes? How many of those have planetary systems? Innumerable ones. You're not alone, you know. With each planetary system comes a unique race, and they're all different, and they're all beautiful in their own part of evolution.

In other words, I thought you ought to know why you can't know everything. And for those of you who are worthy, you are going to have a wonderful surprise, because you have already been a part of it, and you will be made more conscious of that part in the days to come, after reading this. (That's what the radical few of God is all about!)

So get comfortable and read. And in the quiet of reading and digesting what is given here, the knowledge will come to you and depending upon the quality of your patience and your attentiveness, you will gain enormous understanding from my spectrum. And this understanding will free you from a lot of your devoted ignorance, and from your fears which come from ignorance. It will allow you to have your power and be solid in that power, regardless of what happens above your heads. And believe you me, a lot is going to happen above your heads.

The purpose of this book is to engage knowledge. And all you have to do is turn the page and read. And when you get tired, put it down and go rest, and consider what you have read. And when your brain is ready to be attentive, pick it back up and read again.

Here's to your understanding!

It's a tribute to you to know that you
also come from the stars.
From fantastic gods, whose light and beauty
are genetically breathtaking.

Chapter 1

Translating outside the time flow for linear minds. The Ones From Beyond the North Star.

Before we begin, we have to address a basic communication problem.

I am going to have to explain this information to you in a linear fashion, starting at point A and going to point B, because that's the way your consciousness works. I must start in one place, and move forward from there. Your mind, your consciousness, has always worked that way. You come from the past and head for the future. That's how you understand things, like the process of *becoming*. In your mind, you are always *going* to *become* something, and it's difficult for you to realize that you already are.

This presents a serious dilemma in delivering this knowledge, because this is not really linear knowledge. It is absolutely "now" knowledge. It is quite difficult for me to refer to these entities I will be speaking of as past indignations, past creations for a future purpose. Because, you see, they don't live in the time flow, and thus, their knowledge is not created from a past to a future. But in relating it to you, it would be very difficult for you to understand how they are without seeing them in a linear form. So I will endeavor to talk to you "linearly." I will address the past, the present, and the future of their purpose, the main army purpose.

Three Main Groups

To begin with there are interstellar beings, who work with light; there are interdimensional beings; and there are simply those who are "here and now" entities and their craft.

And, yes, there is an interdimensional craft that exists in another time frame, in another vibration. And it does materialize out of nowhere, here into this dimension. That craft is needed because in this dimension, that has everything to do with time, distance, space, and linear form, that craft is necessary for propelling distance.

The interstellar entities are coming from the Seven Sisters, which is termed the Pleiadian system. There is a greater force of entities coming from the Andromeda. These are interstellar entities who go in light. And yet their craft has the ability to travel in hyperlight, which is beyond what light is conceived to be in your physics. As a result, they can be here from an entirely different star system in a matter of moments.

And then there is the group that lives here. And the group that lives here is simply the group that is stationed in the interior of your earth and your government is very much aware of them.

We are working with three different levels of beings who all have their own consciousness and their own interdimensional thinking. But each of these groups have different purposes.

The ones from beyond the North Star.

The group that we are going to talk about in particular are the ones from beyond the North Star. And the ones from beyond the North Star are the Pleiadian and the Andromedan.

The ones from Andromeda, although they are interstellar, are also interdimensional.

Those are my people.

And they are very beautiful. They are entities that are lithely built. They are very thin. They are illuminated in light. They are very tall. They tower to 8' and 10' feet in your understanding of height and look like giants.

They are.

In mythology they were referred to as the Winged Gods. Those are my people.

They do not live on food stuffs. Their biological system is entirely different from your genetic system, which is based on negative/positive polarized energy that creates life forms. Theirs is based on a magnetism that is made up of light. These entities display an embodiment that is magnetized in light.

They do not eat. They gain knowledge. They live off prana. Prana is simply an ancient term that means truth; in other words, they live off truth. They have very large heads, but they are exquisitely beautiful.

In the past, when they appeared, they were called Angels. Gabriel is from these people. Michael is from these people. These, in particular, are the entities that are working with the Ram. They are the White Light. Their craft is illuminated greater than ten thousand suns. They are penetrating in thought. They can reach you in thought in a moment. They have the ability to be stationary in a very large object in their dimension, and at the same time, look down through vibration and watch you. For no matter where they are, they observe you.

Why is this strain so important? Because they are the ones who reseeded this planet when it was destroyed. Your "civilization" has been destroyed more than you know. Once by a very careless civilization who poisoned the atmosphere by releasing atomic power. This group has also reseeded this planet when a civilization collapsed in consciousness, due to decadence. They are the ones who have reseeded to bring about this race that you are looking at right now. You, genetically.

So, in your genes—genes, not your spirit—you possess the genes not only of me, but of the gods. Literally, within all of you, you possess the seeds of Gabriel, and of Michael, and of the Ram. Your appearance is more given to these people in the Andromeda and the interdimensionals than any other civilization inhabiting forever. You look more like them than others that we will talk about later.

Now, why from this strain and from the House of the Ram? Because these entities are more closely aligned to the original physical makeup of the genetic human being. When I originally told you the story of creation, I talked only of the creation of your earth, Terra. I did not talk to you of the creation of other galaxies and civilizations that have lived for billions of years prior to Terra. I gave you a linear description of man and woman, of Deborah/Duval, Duval/Deborah, and the seed of the human being in this atmosphere, in this galaxy, on this pitiful speck of dust on the outer parts of the Milky Way with the yellow sun. I did not tell you of all who lived before you, genetically.

In the beginning on this plane, you really did look like apes. You all did. You were hairy, bent over, and bow-legged. You smelled bad! It was the beginning of the human being. It was the beginning of the god in the experience of reality in this dimension on this plane. It was the beginning of evolution in this understanding.

During the time of your evolution, the gods—those that lived in other dimensions—impregnated the human being, seeding you with excellence, continuously upgrading the physical being, upgrading your ability for physical experience. And that is the precise reason that you look similar, though not altogether the same, as they do.

These caretakers, these angels, if you will, are not angels because they have naive ignorance, but because they are a Lord of Hosts. They have the ability, the understanding, the knowledge, the longevity, and everything else necessary to impregnate and seed the human species to cultivate its continuum.

Yes, they are angels. But they, like you, have highly advanced souls. They possess a soul and spirit, but there is no *image* that binds their spirit. That is why they are illuminated.

Are they evolving? Absolutely! But they are not in a linear process, as you are. They are, instead, in a process of discovering how to take the soul, turn it inside out, and live it.

They are truly beautiful beyond description. You shadow fragments of that beauty. They have taken care of you because you really are their brothers and sisters. You all came from the same source. You are all created in light. You all possess it; it is the common bond between you. It's just that their evolution very rapidly became an involution. In involution, they went through the seven understandings and back again very rapidly. Where they are, in your understanding, is sort of like what you think the seventh level would be. That's where they are.

These beings are my brothers, and they are also genetically part of each of you. They have been your overseers for civilization upon civilization. They are threaded through the mythology that has survived some civilizations. In fact, their comings and goings have been rather astoundingly recorded.

They are responsible for the children being led out of Egypt. They are responsible for Buddha climbing the wall. They are responsible for Mohammed and Gabriel's sortie. They are responsible for human evolution and for keeping it from collapsing upon itself.

Not too many of you have seen them. But in the days to come they are going to be seen. And not too many of you could look upon their countenance and not be affected forever by it. In fact, not too many of you could look upon them—what is the end of your sentence—period.

Why have they sort of protected you from your own evolution? Because there have been times in your civilization that the process of your evolution has been held by a thin fragment from the end, so powerful are you in consciousness that you have the ability to collapse that consciousness upon yourself and end awareness. They have interceded, over, and over, and over again, continuously pulling you out of your loins and your wombs, continuously pulling you away from your single-minded power, your tyranny, continuously pulling you away from your absolute insistence on being ignorant! They have pulled you away from annihilating yourselves over and over again. You are often stupid people! They have kept you from destroying the organism that this earth is, the emerald of your universe.

Why? Because they love you. Because that which constitutes you, outside of this body that you wear, is a great spirit and a soul. And that spirit connects you to forever, and ever, and ever. And it is the love that binds that keeps you going forward. That's why they do it. I cannot begin to name to you the civilizations that have come and gone. Many civilizations have lived here, many of them much more advanced than you. But they have destroyed themselves through decadence, and they have destroyed themselves through power, through releasing the atom and breaking up matter. Their bones and remnants and memories lie at the bottom of the ocean or they are in the process of being refined as molten rock.

And how many people do you suppose were salvaged from those incidents of absolute destruction? Are all of you the seeds of those few survivors? How can a whole civilization be put asunder and yet another one live?

Such survival has only been possible because you have big brothers and sisters who have taken care to make sure you go forward, to make sure you don't collapse upon yourself. They have been your caretakers and your guardians for a long time.

They are the ones responsible for the green explosions in your stratosphere that continuously neutralize the poisons you keep spitting into your atmosphere. If they hadn't been doing that for a very long time, this earth would be shrouded in a black cloud and there wouldn't be any life left here. You insist on polluting the air. What is it you tell yourselves? "It'll go away. It'll go somewhere. Maybe it goes behind the sun." It doesn't go anywhere! It envelopes this life.

These big brothers and sisters have been responsible for a few remarkable human beings. They are, in part, responsible for setting up the ancient schools of wisdom. They were the teachers in the ancient schools of wisdom. They were responsible for kidnapping, seducing away, certain individuals. They were responsible for opening these individuals up, giving them the waters of life, and returning them back to civilization to teach you the truth.

Now, you might ask, "Why don't they simply tell us what we need to know and make us institute change?"

This is where the linear difficulty comes in. Why couldn't I take all your brains out, throw them in the water, heat the water to boiling, and burn away all the ignorance? Why couldn't I just drill a hole in your heads and send a wind through to clean you out? Why couldn't I just plug you in, turn you on, and play intelligence to you? Why couldn't I just do that for you, and save you the trouble and me the frustration?

Because there is one great law and that is the law of the will of expression.

It is termed free will.

It is an ancient understanding that you are brothers and you have to evolve according to your will. It is respected. Free will must evolve in its own understanding. That's why your brothers in space stay hidden in the clouds far away. They send you seeds of truth that will inspire and aspire you to greatness, to start you thinking like an individual. Because that allows you to evolve to where they are. Then you can communicate.

If they appear and invade you, what would you have then? You'd have one bunch of ignorant people who superstitiously believe in the resurrection of God coming to save the world, who would throw themselves down and bow. Another group would run in terror and hide in the mountains. You do not have knowledge that way, nor do you have evolution.

Those who have superstitious and fatalistic beliefs lose the ability to think when they are intimidated by greatness. They too will ultimately grow, but for now they are no more than slaves to dogma and fear!

And what exactly is the underlying fear of those who would run from the lights and hide in the mountains? Do you think they will boil you and serve you up? What would you be, human pilaf? Pilaf! Do you think they would stuff a pumpkin in your mouth? Those bizarre and inaccurate perceptions are precisely why you don't have global communications open right now. That kind of mind-set, that way of irrational thinking is why you have always been intervened upon. You have needed it. Throughout your history this group, this great illustrious group, has intercoursed your women and seduced your men. And they have created great people from their seeds. You are those seeds.

Each and every one of you carry within your being a natural longing to look into the sky at night. That desire is actually genetic. The spirit longs for nothing but God. The body longs for its genetic tree.

Throughout your understanding of time, civilizations have been destroyed and their souls and spirits have reincarnated. My people created the "seed" people, and the seed people are those who survive. And do you know who the seed people were? They were the ones who were warned and told to flee to the mountains or into the caves.

They are the ones who survived everything. Don't you know that your genes are all hooked up to those survivors? And then, to regenerate moral dignity, and spiritual and moral qualifiers and illuminated minds, you have intercoursed with these beings and that created greatness. This greatness ensured a natural destiny, a forward movement, and a natural evolution. And as the evolution progresses, one day the ancient schools will be able to come again and the highways in the sky can be open again, as they once were here on your earth.

The great pyramids that have been ravaged, and the one that was saved, were natural temples of the ancient schools when the highways of light were open through the dimensions. Selected initiates went to those schools. Needless to say, superstition, and myth, and religion, and dogma have gone about destroying these schools throughout the centuries. These schools of illumination have been destroyed so that the world can be kept intact through fear.

But know this. There is coming a day, very soon, when these schools will open up again, and time, distance, and space can be collapsed. And when you collapse the dimensions, my people will be able to interact with you again. But you have to earn the right to be physically alive in this plane when that happens.

This is not about hysterical worship, dogmatic, fall-on-your-face, kiss their feet, and hope for your soulmate responses. (Listen women, just because you can't find a good man, don't think he lives in the Andromeda! And don't you dare ask me if your soulmate is Gabriel. Don't you even dare! Because you feel *soooo* connected. My God! Everyone is connected! A little celestial jesting here.)

The point is, you will be chosen when you have earned the right. Some of you have already been chosen, and some of you have a very good idea who you are. The choices are made for very good reasons. There are "seed" people to illumination. And illumination isn't about riding a ship somewhere. That is a no thing, a side effect of worthiness. And it has nothing to do with sitting down and eating with an alien. That is a phenomenon. That is a side effect. Do you understand?

They would serve you up truth, and you would sit there, waiting for the meat to come. You are different, you know, in the way you think.

The seeds are remarkable people, for they began with simplicity. When I say simple, I mean their minds are uncluttered, while still possessing the capacity for brilliance. The people chosen to be seeds are naturally simple. The more simple they are, the more naturally engendered with courage and tenacity they are. And with courage and tenacity they can live a truth, regardless of whatever test they are put through. The meek are always chosen by my people as the next-step civilization, a greater civilization. We are speaking here of superconsciousness. It is the ability of simple people to be aware in consciousness, to grow, and to be empowered through consciousness, that makes their selection a very natural one. Then perhaps, once again, we can open up the communications, the sharing, the equality.

The more you become involved in consciousness, and learning to direct its energy, the more equalized you become, not only to seen, but to unseen intelligence. That's how you make yourself worthy.

Through this process of selection, by people who are their brothers, selected people become the baby-sitters of time. They will take care to preserve the truth, the reason, the meaning, and the purpose of the human being, so that the truth of what God, the light, is all about can be preserved. Preservation of that truth is what allows a civilization to go all the way to the seventh understanding in consciousness, its seventh realization in evolution.

Inside a simple human being, truth rolls out like a river, and courage and strength are great. There are great human beings, and they are worthy of the gods. Because they are gods, both genetically and spiritually.

Now, those are my people. And their armada is legion.

When I speak to them, they listen. And it isn't as if a king were speaking to an army; it is great god speaking to great god. And there is a harmonious knowingness between us. They know who I am, and what is here that you are not seeing. They know what is happening around you, this moment, that you don't realize. And when I speak, they listen, for we are harmonized as god.

We have done a lot throughout your historical civilization. A lot of you are aware of it. Some of you have given credit where it belongs, through your word in the wind. The recollection of truth has been preserved up until this time. I came back here, bringing a legion with me, for the preservation of truth, to harmonize and equalize God/human being for all the times that are coming.

In a retreat, I anointed some of the people who study with me with an oil that ran into their eyes. I greased up their palms and they became my people. I anointed them, and they are in contact through my house. They live in my consciousness and that is a consciousness of preservation. That is an alien preservation. It is called the gods.

I have reached out to you and secured your readership for a great reason. Truth. You have an opportunity to finish out your evolution, and I'll drive you all the way home. And you're going to have the opportunity to see my people, my beautiful people, and to see their light and their vehicles. They are the Great Intimidators, intimidators of ignorance, that is. And that is right on schedule.

There is a very great ship, called the Mother Ship. It does exist. It slides invisibly by night, because its exterior has been created to be invisible, simply as a result of the metal that it's made out of it. It looks black. But when it passes in front of the moon, it blocks out the entire Eastern portion of the sky at night. That ship houses gods. The Great Ones. My people.

If it was to light up, it would be the sun illuminated ten thousand times! And you would not be able to look upon it with these eyes of yours. And it's here in your stratosphere. Moving quietly and stealthily, it has been over much of your country.

Your government is aware of its presence. It's ominous to them! This ship has warriors that would defy your imagination and gods that would defy the term beauty. It is the ship of legend. And it can sting like a scorpion from above. In a moment, it could blast a continent to the bottom of the sea. In a moment, with one of its lights, it could rotate this continent and send it spinning in orbit. In a moment it could take this world of yours and turn it upside down. In a moment!

Now, do you really think your aircraft bombers could affect a ship like that?

This was the ship that blasted Sodom and Gomorrah off the face of the earth. In one blast. And it didn't take anything to do that. And yet, this great guardian is my people. Those are the gods.

They are here for a lot of reasons. I can't tell you some of them, because you're just not ready to understand them yet. I can tell you this. They are here to help you see the truth behind all the things that are coming and to help you understand what is coming.

It is destiny.

And it is beyond your ability to change the destiny of what is coming. And if you knew the whole truth, you would not change it. You would let it come forth, because it has to. It's the working of a greater plan than you, in your linear mind that works from A to Z or 1 to 7, would ever choose. There is not one of you reading this who has that capacity. You are not illuminated. Because you are not beyond fear, worry, pain, and feeling so darned special that you think you know it all.

Do you know there are some of you reading this who I can't even talk to? You know it all! You are the only impossible thing that exists in this galaxy! If you left destiny to the know-it-alls, you would all suffer greatly. Because they don't understand what the point of universal evolution is all about. They don't know what your destiny means. They can't conceive of the potentially glorious ramifications of each of your lives.

A few of you have great contact with these entities. Their number is 32. Their symbol is the triad. There are some of you who will begin to sit and draw triads for no reason; you will simply feel drawn to do it. The number 32 is simply a code. It is also you. It is also contact. It is also preservation. It's also the thing that inspires you to your truth, to be truth, to live truth. It is like a greater source, helping you to burn away your limitations and to understand. It's a code that pushes a button. They are your brothers, and some of you have already been touched.

And there are others reading this that have been touched by other aliens, and we'll get to that further on.

This great ship is called Melia Unmuun. It means the Silver Life, in your terms, if you were to attempt to define it within the framework of your vocabulary. It is already arranged for some of you to see it, because I want you to look upon power, be in awe, and know you are connected with it. And if you can do that—look in the face of power, be awed, and know you are connected to it—that will do a lot to inspire you on in consciousness.

This ship is here partly because of the disease of the planet and partly because of the collapse in consciousness of civilization on this planet. It is here also because of the changes that are going to occur on this planet and also because of another force that is engaged here on this earth.

This was the great ship that appeared at Fatima. It was the great stinging, burning sun that dried the earth from the rains that fell when your atmosphere fell onto the earth. It was the vision that gave truth about the wars that would come and about the fall of the Great Whore, which is the Catholic Church.

That ship has also inspired other prophets in the past. It has inspired humankind to do wonderful things, for the purpose of preserving seeds of people. Its truth and its words have been misused, misconstrued, and used by powerful men to enslave people. And it's been allowed. But the ship is back for truth. And I'm its prophet.

When I told you to look through the window of change, as I did in a book called *Last Waltz of the Tyrants*, it took a lot of courage for you to do that. It took a lot of courage to change your life, for a day that is coming that will be a signaling of a decade of tumultuous change. The majority who heard me say these words when I delivered them to an audience before they became a book, changed their lives and moved their homes. They went about making changes in their lives, even when it meant they had to be ridiculed. And let me tell you, those changes are paramount to survival.

The information that was given was no different than what has been told to other people of other civilizations that you are the seeds of. It was information that told of change, and of time, and of things that are happening.

The comprehension of that information and the willingness to act on it was one of the things that really determined the worth of many people. The degree of contamination from social consciousness really determined where the teachings went after that time.

Can you act on information? Can you make changing your life a priority? Can you transcend social consciousness, what people think and say about you? Can you understand that time is an illusion that is fickle and changing? Can you be aware of changing times?

These people did that, though the world laughed at them. And many of them splintered off, and went and found teachers who told them "It's all right, you don't have to do anything." And some of them have come back, because what I am telling you is the truth.

Many of you selected yourselves by the great decree known as *taking action*. And by doing so, you, and I, and this ship are bound in destiny.

You were encouraged into taking action in your life for the purpose of getting you situated, to get you busy growing in consciousness. You needed to know that you had a lot in your root cellars and pantries before you felt secure enough to give your full attention to learning. People who were fastidious in doing their "home" work—people who got their house in order—were then free to come and learn with me, without being worried about the concerns of personal survival. (You can't teach a fearful, hungry human!)

There are some individuals who heard that teaching and chose to play trick consciousness. You know what trick consciousness is? They told themselves, "That's enslaving information, Ramtha! I'll survive." But the trick they played is on themselves. There is a whole purpose in delivering that information and in the process of absorbing and responding to it. There is nothing "up there" in your skies that is going to suck you up in a ship and carry you away. That is not going to be the case here.

I am trying to get you through an age. The purpose of this information is to see you exalted in consciousness, all the way to superconsciousness. The purpose is to preserve you through an age, through a change. You have had the most powerful advanced declaration of information and you have probably heeded it better, even thus far, than any other civilization has to date.

I must admit it has demanded really getting to know you comic characters. It has demanded seeing how you live, so that I can explain this information without creating terror or having you set up a whole new dogma!

I have simply wanted you to get your house in order, so that you could sit and learn and not be worried about preparation. I have wanted you to know that your physical needs were taken care of, so that you could be free in consciousness to learn. That is what I have wanted of you, and for you.

We are here to set up a continuation.

These great people — my people — are here to set up, once again, a continuation. But this time it will be an exalted one, without complete collapse of a civilized people, and it will set you into motion for the next ten years of change. All the while you can be fundamentally taken care of, free to grow in consciousness, and the great lost mystery schools can open up again. And that is what I am here to do. I want to. Because I love you.

In my school they are going to learn, learn, learn. They will open, open, open. Regardless of what the rest of the world is doing, how it's shifting, wobbling, moving, dying; in my school, they will learn. And a consciousness is going to grow from the people in this school. And they won't have to go out in the world anymore and work and move about. They will stay put.

There are many people migrating to the Pacific Northwest who've never heard anything about me. They are simply moved by spirit, and it's my spirit. This truth doesn't fit in their belief system. If it did, they wouldn't be moving. So, for them, it's trick time. You know about tricks! You create an illusion. You appear as a woman who's divine and everyone automatically recognizes you as Mary. (You, of course, don't tell them any different.) It's just that, well, that's their image of you. It is simply that through that image you can tell them something and they believe you! That trick has been going on for a very long time. We've always had to send the image of Mary or some other hallowed figure to get your attention.

So there are people coming to my area, the Great Pacific Northwest, who have been inspired a different way. But nonetheless, they are coming here. They just knew they had to be here. They are being drawn.

There are people who are moving out of the cities and they don't know why, but they feel a real urgency to move out. Some of them are having interactions with ships in the wilderness. These people don't even understand it, but they are moving anyway. They don't understand why, but they have some innate sense to put up provisions. And it doesn't make any sense to them really. They don't go around talking about it at the office or over lunch; they are just doing it. They are moved by the spirit. It's the same thing, but the desire is more acceptable if it is internally created by a sense of instinct. For them, that is more identifiable. It's all image, you know?

People are migrating in many directions toward safety. They are not necessarily the Ram's people, following the word, but they are people of spirit and that is the same. They are moving for preservation.

Your brothers and sisters are endeavoring to engage you to get you on with evolution, so that you evolve in knowledge, and in truth, and continue to swell this consciousness. And they are endeavoring to get you to live your truth, so that it becomes the living word. When you have taken care of your personal business—which frees you to learn—that is when the bridging will occur. You will see wonderful things and experience things you cannot comprehend.

These people of mine are not your raiders. They have been your savior. And unfortunately, they have been called messiahs. The intention of these people you called messiahs was simply to provide you with the truth. So that you might be illuminated. So that one day you could join them, and be with them, and not be separate from them.

You're separated now because you are separated in ignorance and still tinged with superstition.

Hysterical people are very superstitious. Dogmatic people who consider themselves very religious are looking for something to worship. The ones who are going to have this contact are the rational, the intuitive, the mystical, the gods. They are the ones who are making major changes in their lives. They are the ones who are living their truth. They are illuminating it. They are the ones who will sit down with my people. Those who will no longer be separate from consciousness, and their God, are the ones that are going to harmonize with them and equalize with them. That is the way it has always been.

If you need to categorize these entities or call them something, you may say they come from the House of the Ram. Because the Ram also is the word that is existent with the triad.

It's wonderful to realize that you have seeds
in the universe.
And perhaps the greatest tribute
to your seed is that it is somewhere else,
living its life happily.

Chapter 2

Interstellar Entities.

Some of you have had some nasty experiences with these people. Or, I should say, you perceive the experience as nasty because your experience has only been as great as your fear of them. And as a result of this experience, some of you have little implants, devices, sitting right beside your optic nerve. And some of you have those implants sitting up your rectum, right near the mouth of the stomach, where it passes into the intestines. Now, that certainly can be a nasty experience.

Those of you who have engaged that group of beings and have that probe sitting within your body, have engaged a group of entities that are interstellar entities. They come from within your Milky Way, from a place where their sun is blue, not yellow. They are also your brothers, and the place that they come from is so very old. These beings never die. They are here to also seed, but they are here to take the seed home.

They are not mean, nasty, or cruel. They are decadent. They are humanoids who possess a soul and spirit. They are in the throes of the dying evolution, for they have become so intellectual that they have burned away emotion. They have great difficulty in understanding love. Touch. Embrace. A warm hand. They are the galactic geniuses, but they are poor in spirit.

These are the beings that are the most notoriously known. Their primary leader is a woman who is beautiful, to a degree of understanding. She is very strong. She comes from the people that you are connected with genetically. She is a part of their strain, and she looks different from all of her people. She is, in a sense, their goddess, if you wish to see it that way. She is endeavoring to reproduce the strain of her genetic line, which is actually most similar to yours in its place in the process of evolution. In other words, civilization is attempting to combine and couple the civilized scientific mind with the spirit, the intuitive, the spiritual, the mystical. That is the idealism they are endeavoring to replant and reseed.

Some of you have a latent memory of an experience with these people. It was too terrible for you to rationalize and remember because you see people in your own image, just like you created God in your own image. And when you see anything other than that image, you have been conditioned to be fearful, superstitious, and even angry. Because anything that looks different is abnormal. Your society has conditioned you to think that way. You've seen too many horror movies!

You do this to your own society every single day. You have human beings who live among you but are outcasts because they don't look like you! They are missing an arm. They have no hair. They have sores over their bodies. They are crippled. You are so conditioned to your own image, you can't bear to look at them without dismay!

These beings I am speaking of here don't look like you at all. This woman's people are small and very fragile, very tender. Their bones are easily broken. If you grabbed them, you would break their limbs and their great black eyes would fill with red water. They are very thin. Their skin is not bronze; it is not burned to death by laying under a yellow sun like yours is. It's not leathery. It's not white. It's not black. It has a bluish tint, consistent with their sun. It has a grey hue to it, therefore they have the appearance of being unhealthy. They are very healthy, within the understanding of their genetic capability.

They have, however, over billions of years in the time flow, eroded away genetically, and are no longer great and beautiful like my people. In fact, through evolution they have genetically become the outpicturing of exactly what they were evolving toward: scientific knowledge. Knowledge. Knowledge. Knowledge.

At one point in their genetic history they had a very firm belief based on doing away with the emotional, or what they took to be the weak side of character. To them the emotional *was* the weak side. As a result, they have literally bred out emotion and bred in intellect.

They are geniuses. They are the ones who can set a nova into motion and explode it. They are the ones that are the "matter gatherers" of the universe. They gather matter. They create from it. Matter creates in front of them. It is co-existent with their image. They can create their image out of raw matter.

I know this sounds bizarre to you. This kind of information is hard for you to take because you are conditioned to think in fear terms, in monstrous terms. You don't have the mind to think of a civilization that is that powerful, yet that empty.

So, what have they done with some of you?

They are the raiders from above.

And, yes, they have created specimens much like those in a laboratory.

They can have control over you because their power of thought is quintessential. They can pick up matter and assimilate it. They can hypnotize you in a moment and distort your time. They can put you, and everything in the surrounding area, to sleep with thought. That's how powerful they are.

They have been interbreeding with you for a very long time.

They have kidnapped the human species and endeavored to take human beings to their world, to breed them. Yes, breed them. But the human being cannot live in their environment because the human being must be impacted with electromagnetic gel through the intestines, through the organs, and through the senses in order to survive the journey. So taking you to them doesn't work, and now they have come for you here. They make their selections on the merit of your genetic ability to be emotional. You are selected on your ability to be warm and loving.

If you are an emotional person, you may well have been selected. They select you on physical endurance and emotional ability as well. When you love, you are numbered. When you feel, you are numbered.

This kidnapping process has already been written about, but not everything could be told, because the world, unfortunately, is not ready to hear all the truth.

It's not as bad as some people have made it out to be. Yes, you share your seed. You are copulated with. You share your seed with your brothers. Yes, it is without love. It is without passion. It is even without gentle emotion. It is mere copulation.

Your seed is extrapolated from you. It is genetically seeded with their own, and the child is raised in their environment.

Your seeds are reseeding.

These people have been called kidnappers. They are despised in some circles and feared in some circles.

And the populace, at large, thinks it is completely ludicrous. They think it's the wild fantasy of a bunch of neurotic people wanting a little publicity.

Don't you know that is the way your civilization has always destroyed the facts they didn't want to look at?

The truth is, these abductions have happened on a very large scale in your population. There are many, many women, with this book in their hand right now, who have been kidnapped. They have been copulated with, and their child has been taken.

You have children in another world.

But those children belong to that world. They are reseeding the human emotion. Those great luminous eyes of theirs are the eyes of forever, enameled in black. Their goddess was genetically created through just such science.

These people all look alike. Their skin tones are all the same and their eyes, those enameled black eyes of forever, have all been the same. But now there are children there who have great eyes that are blue like the sky, and they weep tears that are paler blue than their eyes. And the beings think the tears are wonderful. They collect them! It is wonderful. It is the water of life being regenerated.

Their way of handling you is the only way they know. They believe the human beings they have selected to breed with are superior enough genetically and emotionally that they will draw to them, as offspring, souls and spirits that will pick up evolution at a different page of the book, and begin to evolve and copulate and create love, which is gone from their world.

So some of you have children in the stars who have great, luminous blue eyes, or great black ones. They have small muscle tone for they are still frail. But when those children are bred again and give forth their children, that line will begin to evolve. And that's the purpose behind it all.

If you are reading this material and you are deeply affected by the emotion of what you have read and you know in your soul that this has happened to you, when you are ready—ready in consciousness and maturity—you are going to see an image that will pass through the dimensions to you from that star system. You are going to see your children. So be it.

The phrase "when you are ready" means when you understand that you are not decadent for having this happen to you. And you are ready when you're not sitting there thinking about ejaculating yourself into the vagina of an alien, or making love to an alien. You are ready when you are sitting there thinking about love, the consciousness of the life force, and the fluidity of the life force.

You will grow in consciousness to understand who you are. You will grow in consciousness to understand your connection with the IS. And when you are that mature, you are going to see the images of these little children, big children. And at that point in your maturity, what you see will not be large luminous eyes coming to get you. Instead, you will see love emanating from those globes. That will be a spectacular gift.

This is already set into motion because those who have done all of the raiding from above have learned something from this genetic sharing. They have learned about a mother who misses a child and a son being longed for by his father. They understand now that your soul knows when a seed part of you is gone somewhere else. You see, part of what these people have to learn about is emotional attachment and free will. We all have our learning and that is their learning.

Now, this kidnapping has happened on your plane in very large numbers. But you must understand that this group of entities—ETs you call them, for you have made them famous by showing their image around the world—are purely thought. They operate on orders, thought. They have no emotions. They never apologize. When their ship comes after you, you won't remember it. And if your nose bleeds a lot, it's because they got you. They have a connection with you, and they will for the rest of your life.

Their understanding of love is a thought. When you see their face in front of your face, they have connected with you. They have taken your seed; they have engaged your womb; and you have parented genetically their hopes for forever.

You are sitting there, reading this, thinking this is unbelievable. Believe it! It is so. There is more to the understanding of these people and their relationship with you, but I will only say this. You and that face will be connected for the rest of this life and beyond. You are sort of married to them, in a sense of what you understand marriage to be.

You have partners in the universe. You have partners in life. Consider them a member of your family, because they are.

Would you like to know how you've been acquired for the purpose of raiding? Some of you are taken out of your bed in the middle of the night. Some of you have had your car stopped in the middle of the way that is free. When I first delivered this information, there was someone in the audience who was raided at a fair and one was in an auto machine on the freeway. In a moment, one was taken, engaged, and put back. And they thought they only had a flat tire, along with a sore nose and an ache in the penis, or in the womb, whichever the case.

It doesn't matter where you are. These beings have a way of sterilizing and paralyzing time. They can place an image suggestion in the people around you, so that they never even see what is happening. The person standing right next to you might have no idea what is going on.

Your husband can be asleep beside you and never know that you were taken out of your bed. You could be sleeping right beside your lover, and they might never know!

And you thought the blood on the pillow was only a nosebleed in the night.

Many of you women have had miscarriages, but not all miscarriages are because nature had to get rid of the fetus. Nor are all miscarriages because the spirit rejected the child. Some have occurred because the seeding was not from your husband or your lover.

If you think back clearly to your carriages that you missed, you will find that perhaps within the time flow of conception, you experienced nose bleeds, or perhaps you experienced time loss, or nightmares and dreams of ill fate. If you put all this together, then you will understand what really happened. Those of you who symbolically continue to see or dream of animal forms, that could be occurring because that is the image they pressed to you so that you could accept the way they look.

If you think that your spiritual deity is an eagle, it is probably an alien. Why do you think the Indians saw the eagle as an exalted spirit? Why do you think the tribes saw animals as exalted spirits? Because the tribes were connected with the people in the stars, and their memory of the Great Spirit is the Great White Buffalo, the Great Eagle, the Bear, the Cry of a Killer Whale. That's what those images, in essence, were all about. They were really contact from the stars. It's a great truth.

Now, you men. If you're attracted to dark-haired women with large black eyes, here is a possible reason. If you are prone to fantasies about such women, more than likely it is simply a memory that keeps playing and playing in your mind, especially if this fantasy has to do with simply an erection and a spilling of your seed beyond your control. And if your penis has gotten sore, or if your loins have gotten tender, and if there has been a disconnection from time, and you have suffered with nosebleeds and scratches in the skin, you have had a contact.

To the extent of your ability to remember, I will help. But not if you're going to be a hysterical star mother or claim that surely this entity is your soulmate! But if you are earnest in acquiring knowledge, I will help you to be enlightened regarding these matters.

If you don't have the foggiest notion of any of these things, if this doesn't ring any bells, then more than likely nothing has happened to you. And perhaps that could be termed as a blessing. Or as a longing. It could make you happy that you've never been touched. Or it could make you sad that you were never chosen. Either way, it's your reality.

Here in your culture, at this part of your evolution, your copulation is a very sacred thing. It should be. For my people, it was a very sacred thing. It was not simply lust; it was engaging the light, engaging the spirit. When you struggle as the unified one, that is explosive creation in light!

My people that I have also spoken of in this text have also shared your seed, but they did it as gods who chased the nymphs through the forest. They have been the great lovers of history. They have been the Great Gods. They understand passion and what it is to create in passion. They did not engage you as abstractly as these other beings have.

Some of you, genetically, have had interludes in past lives with the gods. None of you have had interludes in this lifetime. And the likelihood of having one is nil.

By the way, just to show you what is possible within this realm, there were five people in my audience the day I delivered this information who had gone out of the universe and come back within their own capability. Five who knew all there is and had only just a small veil in front of them. Those five have traveled on light and seen the colliding of dimensions. They have gone through a black hole and been compacted in coagulated thought. They have actually ridden the light and looked upon the face of beauty. They understand why one being was called a god, because they looked upon his face.

Those five have made an immaculate journey, and they came back again. And there is a reason for that as well. They are meek. They were pure in spirit. There are five of you who know exactly what is going on behind this veil when it gets so thin. The moment that it becomes transparent, they know. Then the veil clouds over again, and they forget. But there are actually moments when they have a vision and they know what a fiery chariot is all about.

They know who divinely intervened and sat on the Mount of Olives and created thunder. They saw who Moses looked upon. They saw who Jeshua ben Joseph went to. And they looked upon the face of a god, the five of them. They didn't give their seed. They didn't copulate. They just experienced.

They were selected on the merit of simplicity. They were pure in spirit—meaning there was no duality. They had compassion, tenderness, and kindness, and they were without a superstitious mind. They could look upon greatness and be in awe of it, but want to be just like it, not worship it, just naturally be "like it."

Those five have wonderful things coming to them, personally and intimately, a parlay, a sorte.

The truth isn't told to everyone, because
not everyone is ready to hear it.

The Book of Books is full of imagery
perpetrated upon humanity just to
get the message across.

Chapter 3

Evolved Tyrants

There is also another group of entities who are best described as evolved tyrants. And these tyrants come from a far distance. They also are based here on your planet. These entities are best described as evolved tyrants, technically. They are beautiful gods, but they are "evolved tyrants."

They were the ones who helped to create religion here. They created a division between people by introducing the concept of worship and the enslavement of the seed people. They are the ones that created the concept of heaven and hell. They are the ones who created Lucifer. They are the ones who created Jehovah.

They are entities that have set the world against itself. They have selected slaves and those slaves are willing to destroy the world in the name of God.

These entities are very powerful.

They have a way of keeping the myth alive, and they have a way of keeping superstitious people ignorant. They have had their pawns in high places, and they continue to do so this very day. Their agreement runs out with the last Pope. They are, indeed, entities who endeavored to seed their truth, not with enlightenment, but with subservience.

They are the slave drivers.

They've had their time here throughout the ages, and they are participating in other civilizations as well; they are not just here. They are not interested in taking anything of you genetically; they are quite exquisite in their forms. So they don't need you. They need you only to govern.

They, too, are doing a righteous thing *within their consciousness*, because one part of them did help to uplift and evolve some ignorant human beings. When I use the word ignorance in this instance, I am speaking in terms of evolution. They helped inspire and aspire the first dynasties in Mongolia. They also helped to inspire and helped to create the dynasty of what you call ancient Egypt. They are technically brilliant entities.

They have endeavored to build up the human condition, but only good enough to serve them. Do you understand? That is how most unenlightened parents raise children. They raise the children to make *the parents* proud. Get it? But the parents always *own* the children. Those are "their" children. They make "me" proud. They make "me" look good. They are "my" family. That is the consciousness of these entities.

The concept of "my" family is a tyrannical attitude, seeded from tyrants who see you as "their" people "they" created. Because they instilled sterile civilizations, they have instilled a concept of subservience, ignorance, and enslavement.

They have boldly said, "You will have no other gods before us." Because they are jealous gods...and they are.

They have resurrected saints to pass on their truth and they have resurrected demons to keep it intact. Then they told you the demons were the aggressive dark side, supposedly, of the human being.

And it's all simply a reflection of their need to have you serve them. If you really knew the extent these entities have played in religion, down to even creating major figures in religion, some of you who are tenderhearted would literally find the information fatal. It would leave great gaps in your soul. There are no words I can call upon to tell you how influential they have been in enslaving you to dogma and other things!

Any individual can force subservience on a lower being. No matter how exalted an entity is, he has the ability to lock the minds of lower beings into a stagnated growth pattern, even though the being from a lower pattern of evolution is still of god essence.

These entities I am speaking of now have been responsible for powerful governments. They were responsible for the birth of Julius Caesar. They were responsible for part of Napoleon's life. They were responsible for Hitler. Those are some relatively contemporary figures, but those names are nothing in the whole spectrum of time.

All this control and domination has been to keep intact an idealism, so that you are never allowed to go beyond your ability to serve.

One day there will come a great armada, and it will collide with a greater armada. And this is the vision of the war in the heavens that has been long prophesied. And that war is a war of light beings to light beings. It is the control of your rag-tag environment that lies in the balance. All because of the need of an altered ego.

These gods I am addressing here still have a ways to go in evolution. Know that they are already being penetrated. They are already coming to realizations.

Here's what it's about for them. How much can you conquer until you have to conquer yourself? That is their process, as it has been yours—the conquering of self.

What value could you possibly be to them? You cannot even begin to serve them like they would really like you to. You aren't capable of building ships to their demands. You can't change form. You can't disappear here and reappear over there. You aren't capable of a powerful focused thought; you are just now learning that. You are only toys to them!

The "greatest" war that was ever fought was the greatest tyrant against the greatest tyrant. That has always been the sublime battle. The greatest against the greatest.

You do it all the time in your society, and you think it's wonderful! You are a reflection of them and their consciousness. That consciousness is reflected in the image. But their determination is to have dominion over you. And what greater esteem than to own a group of gods!

You have a lot of entities who love you. (Me, in particular.) And there is no greater sword than the sword of truth, enlightenment, and knowledge. And you have in you, the capacity to be awesome entities. Knowledge and truth. The body you have, combined with a consciousness that can fine-tune itself, is far greater than any ship they ride.

The ships they ride, the light that illuminates them, the color of their skin, the sometimes great, sometimes frightening appearance of their being, is all made up of one greater factor. And that greater factor is the IS consciousness. All choices are side effects of evolution in consciousness. Everything I have described thus far is all possible as a way for groups of beings to function in their own process of evolution.

The greatest beings can take the body, just like the one you have, and set its consciousness free. That's power. That's more powerful than that great ship sitting there that could blow this world into turmoil. It is greater and more powerful than any interdimensional craft, weapon, force, or forcefield. Because all of those material creations came from consciousness.

The greatest army that will ever be, if you have to talk in terms of armies, is one of consciousness. Because nothing, nothing, can defy an army of consciousness.

What I teach in my school is something that even those who would take you into submission cannot afford for you to learn. Because when you understand consciousness, when you are aware of what you are, you will never be controlled.

I want you to know that these entities took steps in evolution because of their environment. They are highly technical. They understand matter and antimatter. And they understand gravitation and antigravity. They understand consciousness and subconsciousness, to the degree that it produces antimatter. Don't you know that life that can appear and disappear is antimatter engaged? But what is antimatter? Consciousness.

So they are only as great as their ability to engage matter.

The greatest power comes in consciousness. Consciousness is what constitutes matter.

You have the capacity to learn to expand and ride the light, to get outside of your mind in consciousness. And when you can do that, there is no ship that can track you down. They can't go that fast! They can't find you. Did you know there is not one of you reading this, or one of them, that can find my daughter when her consciousness leaves and I am speaking? Where do you think she is when I am speaking? There is no one who can find her when she is gone, because she goes to the light in consciousness.

Now, let's get back to how you serve these beings I am currently discussing. You "serve" them in multi-faceted ways. The ones I have been speaking of, the ones with the altered ego problem, the jealous ones, are going to be turned around. It's already happened in destiny. You are living in the residue of time; you are still living in the spiral as it is being played out. But it's already happened!

If you act like a victim,
you'll be treated like one.
It's rather like life that way.

Give forth love, and you'll be
met on even ground.
Show them your house
and they'll show you theirs.

Chapter 4

Light Draws Light.
Like Draws Like.

I know this sounds incredulous. Some of you will believe you've been contacted; you can feel the ring of truth to that. And some of you know you are of more value than dust to dust returneth; you can accept that you are more valuable than your dung and your urine. But is it possible for you to believe that no one really wants to scoop you up in a big shovel and take you to another universe? You'd make a terrible human pilaf!

These entities devour truth and knowledge. They don't devour flesh. The little ones with the big eyes that are after you don't eat. They have evolved beyond lips. They don't kiss, so they're not after that! They don't form words, so they don't need lips. They do talk, but communication comes out of their consciousness, not out of their mouth.

They don't wear ear bangles so they don't need ears. (The next generation born of your women will, no doubt, have elongated lobes and very big ears. They will need them to punch holes in. It's the only reason they'll be there, to decorate!)

The little beings don't hear with sound, like you conceive audio resonance. If you could only hear how they hear you. You sound like a very low motor. That's how you sound to them. They don't need to hear you articulate. They hear the vibration that goes behind the sound, so they don't need ears.

They are not into hair, so they don't need it. It only gets in the way. It's time consuming. Hair was created to shelter your brain from the sun and to hide life. Now it is created to do all of these things with; it takes hours of your day to do something with the mess.

Compared to you, the little beings are really superior people. You are simple people, in the process of understanding something, so you make excellent specimens. They understand, and they'll talk to you. Their goddess is quite extraordinary; not what you would call entirely beautiful, but extraordinary nonetheless.

The more engaging you become, the more aware they are. Light draws light. Like draws like. It equalizes itself.

Their ships defy gravity. They can just sit there in the sky. They don't have any wings. They don't flap wings like a bird. These are people capable of creating the illusion that you are still in a room, when you are really with them!

I could tell you a lot of things about how what you call aliens have inspired civilizations, and what they've done with civilizations. But there is only one critical thing I must convey to you. They are your brothers and your sisters. No matter what they look like, they are still god, and within them reigns a soul. They have a great spirit that is consistent with your light.

You just look different, that's all. Your body functions are different. Your needs are different. Your survival is different because your atmosphere is different from theirs. Your basic mineral makeup of physical mass is different from theirs. For the most part, they cannot live in your environment. It doesn't work for them. They do not breathe oxygen.

Some beings can breathe because they are adaptable. They are the Great Gods, the Angels, the Light Beings. They are adaptable because their whole life force is the life force. Their life force is not mineral compounds. It's not gases. It's the *life force*.

You are going to see two different craft in your skies. One will be bloody orange, a rusty blood-red orange. It is that color when it's suspended, but it bursts into brilliant illuminated light when it begins to change form. You are going to see those appear in your skies. That bloody orange craft belongs to the little people. When you look at it, remember the eyes. And remember how fragile they are. If you make a mental picture of them and send it to them in consciousness, you will make contact with them on a friendly basis.

Remember, they're very sterile. They don't laugh and they don't tell jokes. There are no comedians in the bunch. They are very precise. They make short order of things because they are short order! They are very businesslike. But, nonetheless, they will make contact with you. They may not land at first. They will circle you in their craft and observe you for a while. But if they think you're for real, they'll make contact with you.

Those are the little ones.

The ship that belongs to my people is awesome, and you may see it. And those beings of light! You will see their light floating. And you may wake up one morning and see a light standing outside your window, looking in at you. Those are my people.

The other major group we're talking about here, the red and green, you are not going to physically see very much of. Those are the controllers of subservience, the Great Tyrants. You won't see them up in your skies.

The explosion of the green fireballs in your stratum will continue. And that is just the process of continuously rarifying and cleansing your atmosphere, a process undertaken by my people.

In the coming time, it will not be uncommon for you to look up and see ships appearing simultaneously beside each other. They will split off, go to different directions, and then seem to fall out of the sky. Or, as you drive down the road you may see a pulsating light.

Those are the only ships you will be seeing here. And you haven't got anything to fear from them except ignorance.

If you act like a victim, you will be treated like a victim.

If you understand consciousness, and know who you are, you will be met on equal terms. If you act like a victim, you will be treated like one. It's rather like life that way.

Truth is truth, no matter what color its skin, no matter how it looks. Truth is truth. Consciousness is consciousness. And those are things that you have learned that are synonymous in their truth as well. Act like a god and you will be treated like one. It's called equalization...equality. Harmonizing magnetism. Act like a victim, you'll be treated like one. It's the way it is.

Give forth love, and you'll be met on even ground. Show them your house, and they'll show you theirs.

At the end of this next decade—these are tumultuous times—you are going to get exposed more and more to what I am talking about. And you should be exposed to it. You need to be. It is being arranged even now. As you get more information about who they are and what they are about, your mind is being opened to possibilities.

In your maturation process, you first learn to lie on your belly and wiggle around. Pretty soon you get a little momentum underneath and you learn how to crawl. And after you learn how to crawl, you start pulling yourself up. And after you do that, you learn how to walk. All of those things are synonymous with a learning experience, and it is the same process with an experience with your brothers. Because that's how it must come; little by little the knowledge grows.

You will be learning to have a connection with these people which is as natural as breathing and you will learn it in doses. You will gain knowledge, and the knowledge will create the reality. The reality will be the door that will open that will allow you to engage. You will engage precisely to your ability to know, not any more than that, except for the five I mentioned earlier. The next contact for those five will be at an even greater level.

Finally, one day, you will all be "broken in," so to speak. You will be "alienized," until all this is normal. And when it is normal, it is reality. Then you can go to school, and have more adventures, and become aware of great technology. We will have a great opportunity as part of this evolution.

When someone says they had a brief encounter, believe them. The problem is your fickle nature. You don't want to believe someone else's experience, only because you didn't have the experience. But I'm telling you, many, many people who are holding this book in their hand right now are going to have experiences. And the wisest and most prudent thing you can do is listen to what your friends tell you, because you can learn a lot through their experience.

This is no understanding for the concept of disbelief. All things exist. If you could compile a book of all of the facts that have been thrown out because they didn't conform to Christian belief, there would never be the dichotomy of belief and disbelief.

You would be wise, if you want to be part of this ongoingness, to find out all you can about this phenomenon. Some of you will have incredulous experiences. Allow them to be incredulous! And if you don't have the words to share with anyone, it's not important. What is important is that you are *learning* from it.

What we've been doing thus far is defining some things for you. In the next section, you are going to learn what your government knows. You need to know that there is an enormous conspiracy at work. Your government has already heard these words I am delivering to you, but you ought to know your brothers in the skies who are waiting to engage you have heard them too.

When you sit outside and look up to the sky, remember the triad is the symbol that connects through the dimensions. If you wish to send an image, send the image of the triad out. It will hasten your encounters. Above everything else, have your thoughts be consistent with your "wants." Then your wants will be consistent with your reality. And your reality harmonizes realities consistent with your wants. Then when you engage, you will be engaged. That is how it works.

It was never important what you did.
It was what you were and are.
And no one will ever love you like
you will love you through
all the veils.

Chapter 5

What Your Government Knows. And What They Want You To Think They Know.

Your government is a very precarious institution. Your government has possessed the knowledge of antigravity since the turn of your century.

Back when horses were still trying to pull around carriages with little effect on the environment, there were scientists, at the turn of the century, who were fully aware of the secrets of a very secret society that understood the alchemy of atomic levels. This society understood magnetism to its fullest degree; they understood $E = MC^2$, the squared light reciprocal. And the most brilliant and astute of these people was an entity called Tesla.

This ingenious man was part of a society of individuals that were groomed to feed technical knowledge and understanding, bit by bit, to civilization. This information was fed to allow civilization to evolve technically. (You've had such brilliant people who've helped humanity to evolve all through the last two thousand years.) The only problem occurs when kings and governments discover such an individual. They confiscate their papers and imprison the individual, to bleed them of their knowledge. That terrible process of government intervention and power-mongering really hasn't changed at all through history.

Your government had aeroships at the turn of the century. They weren't particularly great ones, but they had them. You know those brothers that were Wright on, the entities who created the conventional bird flying through the sky? It was agreed to allow that discovery to come into effect, the "first" aeroship. But this knowledge has been around for a very long time.

If you understand matter and antimatter, and if you understand how matter lines cross and how the vortex of matter is created, then you will understand how to impel through antigravity. That information has always been here. It's what makes up the cosmos. It's what constitutes this earth. Great minds that have had the secret knowledge have passed it on in legends for years.

The governments of your world took more specific interest in this activity during the war that is Number Two, when the tyrant was in Europa.

His legions had developed aeroships that are similar to the ones that you think you would like to see. These ships were borne under the symbol for the zircon, which is the crooked cross. That symbol was supposed to stand for the Illuminati; and the Illuminati, as this tyrant viewed it, was to be the superior race.

This was, and is, a bigoted, enslaving, tyrannical concept. It is prostitution of a greater truth about what the Illuminati really were. The Illuminati was not a white race. They were illuminated people. These illuminated people go back into your history thousands of years. They were the carriers of truth and technology; and it was their job to seed, through civilization, certain amounts of technology, according to the ability of evolution of the collective civilization at the time.

When I speak of this particular civilization, let me clarify what I mean. Referencing your Christian understanding of B.C. and A.D., this civilization began about 5,000 B.C. and runs through now. Generally speaking, this whole civilization, your civilization, has been very retarded and slow in what might be referred to as technical abilities and attributes.The Illuminati was, and is, an extraordinary group of people who are called the Ancients; and they hold the secrets to all of the universe. That information was given to certain individuals. One brilliant entity, Tesla, gave your government the ability to have antigravity. Antigravity is achieved on a grid-line. Creating a vacuum at zero degrees creates antigravity and that can ride a light!

Now that vacuum system, as everyone technically should know, is created with magnets. So, the tyrant of the Second War had ships, for he too had gotten hold of information by making a pledge to the Illuminati that he would take his information and illuminate the world.

Instead what occurred is what happens when truth falls into alignment with one who is not a consciously engaged human being, but wears an image. And what, unfortunately, occurred was that the tyrant had a personal need to control the world, to exalt his altered ego.

So he had to fall.

You do not destroy human beings to gain power. And this was what this entity did in great measure.

Now, your government confiscated very important documents from this tyrant. And they also confiscated great minds from his land and the Land of the Bear, which is called Russia. These great minds your government confiscated were part of the seed people from the Illuminati. They were brought to your country to work on furthering the conquest of gravity.

In the early to mid 1900s, your government already had the ability to transmute matter. They had already done several experiments transporting great iron ships from one port to another.

They even put this into effect in a battleship which was fully loaded with armaments and people. Its belly hung with generators that created a magnetic force, that created the vacuum, that created zero degrees temperature, that would allow it to raise the vibratory rate of the whole ship and the crew, transporting them to another place and time!

They were very successful with it. Unfortunately, the people who were a part of this experiment have been locked away and kept, their minds destroyed, to keep them from talking.

Your government has since developed its own aeroships. Your government has the capacity to defy gravity. Your government has the capacity to travel in a straight line on an energy line which they create. And even though that sounds fantastic, their ability is still very limited. Your government has capacities you don't know about, but they are still limited.

There is more your government has done. They have gone even further. In an effort to perfect their own ships, they have taken, and are holding, your brothers' ships, the ones that have collided and broken into pieces in your stratosphere. They also have the bodies of some of your brothers from another world, bodies they have retrieved from these ships. And through technology and ancient alchemy, they have been able to transmute some of your known metals, those that make up the base of meteorites. They have learned to transmute certain metals into aligning to better perfect their ships.

Your government is very aware that their earth is under siege. And that it is being watched.

Your army denies you the factual information regarding such extraordinary beings. And anyone who has ever had interaction with either the government craft or those from another world has invariably been publicly humiliated.

Now, why would your government want to do this when they are spending all of your money trying to create this shuttle rocket?

You must understand the consciousness. They have always successfully controlled you by having you work collectively for some wonderful goal. And this particular promised wonderful goal will be a collective economic merger in space. They will tell you this will create medicines that will cure cancer and medicines that will cure your AIDS plague. They will tell you that you must work to support their missions because it will allow them to create marvelous synthetics that will make your life sweeter, and more comfortable, while helping the earth. Whatever it takes, they will tell you!

That is how they keep you at work, with your back bent. That is how they keep taxing you, bleeding you for dollars that go into a program that supports all of these people, to keep them gainfully employed, which keeps your government in power, and your focus out of focus!

They are using this goal like a carrot, dangling it in front of you, and all the while they've been to the moon and back in the twinkling of an eye!

What are they going to do with you? How are they going to explain to you that their system is very simple? It doesn't even require a lot of money from you? How are they going to keep you under wraps? How are they going to control you if they can't tax you?

Wake up!

Now, even though your government has learned a few clever tricks, it does not know how to control time. They have had a lot of small projects that just disappeared. Poof!

Let me ask you something. What is a linear person going to do about linear time, when there is no such thing as time? And your government will not be able to own that secret. So their little projects will go to and fro, on a light beam. Back and forth.

Their projects will not make it to the fiery planet any too soon. Your government keeps sending their little projects off into the discovery of deep space. They explode in the twinkling of an eye and are gone. But they won't reappear until two or three centuries later, because they don't know how to control time. And until you understand how to control time, it doesn't really matter how well you can raise the vibratory level and set a craft down someplace else.

Let's look at the predicament. If you've never been beyond the sun, how can you set coordinates to get you there in the first place? If you guess and set coordinates to go beyond the sun and you try out hyperspace in order to get there, your creations will disappear all right. And, in the time flow, they'll be back, two or three centuries from now. And they'll land in your front yard!

Your government flies their craft around your skies, keeping a very good eye on who sees them and who doesn't. There are a lot of innocent people who have had lobotomies. A lot of people who accidentally looked up in the sky have had their lives changed forever. They acquire amnesia; they don't remember much of anything, all because they experienced government projects.

Speaking personally, I'm not very impressed with either your government's equipment or their state of mind in creating the technology. And you must understand, they've only been given enough information to make toys. Oh sure, it's wonderful that they have the ability to transmute matter, in some form. But they can't override "time." And that is their greatest challenge.

So, if you want to be thankful for your government and your tax dollars, be thankful about time. It is the only "time" you should be happy about time!

The Graymen (tyrants who control the world's money, etc.) must think they can ride this thing out on the dark side of the moon if things get rough. They must think they can come back two or three centuries from now, when their treasury bonds won't mean anything. That's an inside joke. You'll have to read some of my other material to understand that comment, and I recommend you do just that.

Your Graymen are very aware of the real ones up in your skies. But the alien ability to cover-up is so powerful that they can convince the observer that they were in a state of hysteria and they didn't see what they thought they saw. They not only know how to do that, they have done it time and time again. The most powerful people in the world have seen a great ship and been in awe of it. And some of the stupid ones have even thought of how they could shoot it down, and they have sent their little aeroships to chase it around.

Well, you can't shoot these ships down because you're not equipped technically to do that yet. You don't have the power. There are secrets that the wise ones never gave out. They still hold onto many. There is great purpose and reason behind these wonderful illuminated few who have been endowed with technical genius and expertise. In essence, they understand magnetism, gravity, and antigravity. They have been very careful how this information is known, and they have worked on your behalf by not allowing all the truth about physics and mathematics to ever be figured out, even by a computer.

The reason they are so cautious is very obvious. The kind of information they hold could be devastating in the hands of a government with a retarded consciousness. And one more time, one more civilization could fall.

Here's what's important to you personally. If you have contact with a government ship, they are going to do something with you. If you make contact with a brother in space, your government will still try to find you, because your government tracks ships. They are very aware of where they are. They track the coordinates, when they can hold onto them. But it's becoming increasingly difficult for your government to track down every person that has an experience because there is so much contact going on now. There are too many people who have seen some phenomenon or had an experience.

When your contacts with your brothers in space occur because you have both purposefully engaged it, that's when you have an extraordinary opportunity to see knowledge at work in its greatest form.

You will see a craft that has no apparent internal engine. You will see something that can just "sit" above your head and not make a sound. Can you imagine experiencing one, riding in one, being a part of one?

Most of you couldn't ride in one because your equilibrium could not tolerate the speed of light. Your point of balance and how your body is physically made up would not allow many of you to tolerate the irrespective laws they live with in flight. They make a right angle degree turn in such a quick move that it looks like a blur. You could only ride in a linear fashion and make big, swooping circles. Or perhaps you could go up. But not too fast. Or down. But not too fast. Because your environment has not tempered you to that lack of restraint. But there will be opportunities to experience a great ship that will make compensations for you and take you very slowly.

Don't get the idea you're going anywhere though. You're not going to be whisked to another galaxy to live there happily forever and ever and ever. That's not heaven.

It is simply that you are only going to progress to a place where you are worthy of having the experience for yourself and so that your children can have those experiences in their future. There's really nothing your government can do about this, nothing at all. Because when these people want to have their way with you, they will have their way with you. And there is nothing that can stop their great ships.

But this isn't, and shouldn't be, a romance with a ship of light. It should be a romance with knowledge. That's what it should be. Your love affair should be with having the knowledge and the technical ability to understand the pertinent laws. You should learn to understand centrifugal force and to understand gravity and antigravity. That's what the romance should be about!

Do you remember the story I've told before about the genie and the wishes? If a genie appeared before you and offered you three wishes, what would you wish for? Would you simply take your three wishes and feel gratified? Or would you be smart enough to wish to be the genie? Ask to be the genie! Then you are not limited to three wishes but you can create whatever you want for as long as you want. Do you understand the difference? When you understand the difference, you will understand why you can experience more on that level of understanding than someone who simply wants to take a free ride on a great ship.

There is one other important aspect to this free ride business. Not all of these ships are safe. You must understand this. Any ship that operates off magnetism in a vacuum creates a degree of hot radiation. And to flesh and blood and the materials you are made of, that isn't always the safest environment for human beings to be around. That is also the reason that the process of landing burns the ground, and the reason that life refuses to grow there after a landing. That is a side effect of their technology, yet they themselves are not affected by it.

You must be mindful that you really don't want all of these ships and their species to come and land in your yard. There are some you wouldn't want in your front yard. You might not live to see the next morning if some came to visit! In your Book of Books it talks about the Ark of the Covenant. It talks about the Mount of Olives and the Thunder of the Lord and those who could not come upon the holy ground or get too close. That was because of the radiation the ship emitted. That ship emitted a lot of radiation. The Ark of the Covenant was simply a connector of radiation and a laser blaster. It was used to defeat armies. Only certain people could carry the Ark, the Nubians, who had the ability to be immune to the radiation that it gave off.

One day when the consciousness of humanity is just right, you'll see a wonderful thing. You'll see a stairway all the way to wonderment. And that one experience will save ten million words of explanation.

Don't be a part-time person.
Be a fulltime god.
That's what it takes to advance and to
move forward, to burn away the
restrictions and the limitations,
to be all that you can be.

Chapter 6

There Is Life Wherever Life Has Conditions And Matter Can Be Formulated.

We've talked about three main groups and those groups are all you really need to be interested in now. There are others that have never made contact here. There are primitive people in stages who are just beginning the journey of evolution in their own form and way. There is life wherever life has conditions and matter can be formulated. Where those conditions exist, the spirit will move and humans will evolve according to their environment. You are evolving according to your environment. There is much more to your fiber than your moralistic attitudes.

Yes, your genetic material has been milked from you. Yes, they have taken your seed and created wonderful human beings, hybrids, so to speak. And if that was what you call a rape, then the purposeful good of that rape was that it created life, and life where it was treasured, not thrown away, as it is here.

Your seeds produce explosive life genetically, and these beings know how to take your genetic seed and implant in the tree of knowledge, certain things that they want that body to have. They know how to genetically pluck out your impurities and replace them with their purities. Within the next four years, one of the Illuminati will give the science of this to one of your physicians. You will be taught how to do that.

Children will be able to come into this plane through their own selection process and determine how they want their genes. Disease won't be here anymore, because it will be plucked out genetically. A pure attitude will be able to come into this plane in a pure body.

Some of you are diseased and are carriers of disease and yet your seeds have still been taken anyway. A cross section of those seeds have been checked and even you have created life elsewhere.

Life is life.

Life. The principal cause. The life force. A man's semen is very much like a running river. The nest of some fabulous bird is very much like the nest of a womb in a woman.

The life force is the life force, and the most moral objective is to create life with it. Here in your society, no one values that life any longer, but there are societies that are hungry for the seed you waste. And they are taking it. And they feel they have every right to take it, because everyone who has the full knowledge and participates in the life force understands that they share the life force. They understand that the light is a gift to all of you.

Do you know what an honor it is to be alive? Even though you are in sluggish, slow, thick, dense, bodies? Do you know what an honor it is to be consciously engaging this density? Consciousness is God. But god without knowing itself is boring; it's a nothing. It's the *IS*, unawakened. It is yet to be everything potentially.

Your life is more than just sitting on some chair, feeding one hole and having it go out the other. Men, your life is more than just spilling your seed and grabbing a breast. Women, it's more than what you wear!

It is life. It is the force. It is what quickens and illuminates and calls forth unknown universes. And the life force, no matter how you look at it, is God. And to be consciously engaging in that is a privilege.

Can you ever imagine yourself going to some other colony and breeding its women and men and treating the offspring like a priceless treasure? Can't fathom yourself doing that, can you?

But think how valuable life must be to others who would go to such an extent to get what you have left over. They risk their own lives, because of your environment; and they have to go through this whole cloak-and-dagger routine, just to gain life. While here on this plane it's religiously suppressed and made ugly. And that religiously suppressed ugliness creates decadence, and decadence creates disease, because everyone hates everyone because of their sexuality.

It's life force. That's what it is.

Can you imagine growing up as a human being in this plane and seeing yourself as more than just an ejaculating, copulating entity? Can you rather imagine seeing yourself as life? Can you imagine deriving pleasure from just being part of life? From just being?

We are not talking about struggling to have an orgasm. We are talking about creation here. Otherwise, who would be interested in this speck of dust at the edge of this universe, with its yellow sun? Why else would anyone care about this earth with only one moon? Why else? It is because of life. What you are genetically is needed other places.

And that is not to say that these beings who are coming here don't have a lot to share. Your brothers in space have always been in contact with the Illuminated Ones, always. They have shared technologies and understandings. And yes, they have crystal balls and they can look into the time spiral, so they know the future.

Do you know the time spiral is already owned, so to speak? The next ten years has already happened; you're just langoring in the past of this time. They understand all of that and they have shared that future knowledge regarding the next ten years with certain qualified individuals on your plane.

One of the most astounding individuals who ever lived was a woman who was part of the secret people. Her labor was making pottery, under the auspices of her husband's name, because women weren't allowed to work publicly.

Since women's work wasn't allowed out in public, she was stowed away in a cave where she made pottery for her husband to sell in the marketplace. Day in and day out she made beautiful urns that would later be treasured in Crete.

This simple woman had no understanding of reading and writing, which was, at that time, created through hieroglyphics. One day a mysterious stranger, a messenger, appeared at the entrance of the cave where she hid, making urns. He was a very tall, eloquent man. He walked in, sat down, and said not a word. Within a moment the cave was illuminated with light. The clay began to dry on the fingers of this poor, wretched woman hiding in her cave. And when the light went out, he shared an herb with her, walked out the entrance of the cave, and disappeared. And from that day forward, she became of the Illuminati. She held within her secrets and knowledge and truth that was given to her because her mind was open. And in a moment of light, all that he knew, he gave to her. And she lived to be a legend.

She continued to make her pottery and she frequently had mysterious visitors who would sit with her and share an herb. And the visitors would walk away and she would make pottery for her husband, whose pottery became so famous in the land of Crete.

That's what's happening to some of you. I'm not saying you are going to start to channel some god from another planet. That's not what I'm communicating here. People are selected on the merit of their great and profound simplicity. This is simple knowledge, but it's the knowledge of a seeding civilization.

These entities who value life more than you do are here to extrapolate it from you. And they are here to sit down and share with you as well, if you are ready to share.

Not all of them are friendly, by any means. Their ships can certainly be deadly. But when it's right, it's right, and you will engage them. It's part of a learning process. Alien entities are responsible for religious dogma. And why would they enslave you to the tyranny of a variety of religious dogmas that have throughout history resulted in massive death, bloodshed, ignorance and even into what has come to be called the Dark Ages? For what purpose would they create religion to enslave you? The only reason some of these angels, as it were, did this was to gather together seed people and give them laws to live by, thereby creating for them a morality. They wanted you to come back to center and remember to respect life. Throughout history, certain people have been picked because they were essentially simple people with a basic understanding and respect for life.

Look at how you were living then. You were raping your children and copulating with cattle, with sheep, with dogs. You have had a civilization that slaughtered millions of exotic birds, just because you thought the tongue was an aphrodisiac. Those are the civilizations that appear, within your history, to have "gone asunder." And in those times of great civil peril, the Great Ones have appeared, many times, and delivered to simple people, a simple truth. They gave you laws to live by. The people got frightened, so they abided by the laws. The Great Ones have always come to teach moral and spiritual responsibility to a collapsing consciousness. That is why they come.

But little minds have carried a simple, beautiful truth forward into religion. They have infused the simple, beautiful truth with dogma and used it to murder the world. That's all dogma has been used for, suppression and killing.

And yet, the intention of delivering that original truth was to keep truth flowing and to keep man from collapsing in on himself and to keep women from dying in consciousness.

There is a level, not too far from this one, in a dimension mythically called the astral level. (That's not a proper term, by the way. There is spirit there.) It is a plane where consciousness is folded up like a great silken cloth. It is folded very tightly; the light is very tight there. There are actually people, just like you, who have had a collapsed consciousness and they are there, unawakened. Because they have created a reality where they have collapsed themselves in personal beliefs. And they are waiting there for a savior. They can't live a physical life here anymore until their savior saves them. And because the savior has to look like the picture in their mind, or the one on the wall of the room right before they died, they are still there. The "imagery" has to be right for them to believe. They cannot expand in consciousness, which is being dead in spirit, because they have collapsed their own house against itself or onto itself. They all felt complete unworthiness, impelling, compelling unworthiness. And, because you really do create your own reality, the law in their reality is that they cannot awaken until they are worthy!

And what have you learned? That worthiness comes from within the entity and it must be experienced, and it must be experienced according to your laws. Now ask me how important consciousness is! That's how important. These people I have just mentioned to you have lived in this level millions of years in the time flow as you know it. They are dead in spirit, meaning they are dead in consciousness.

So Jeshua ben Joseph has to look like the picture. And he has to be "imagized" as the picture looks. The created image of him has to look exactly like that for you to believe; he has to wear the exact same clothes the artist saw. He must have a brown beard and brown hair, even though he didn't. Because if he looked any other way, he would be the devil, impersonating Jeshua ben Joseph. That's the way your mind works.

And let's be honest here, if we want the ignorant masses to do anything, we have got to portray Mary. Do you know the Mother of God? Well, of course no one is going to believe anything if it isn't spoken by a mother! And if we're going to produce a mother, by all means, it should be the Mother of God. That's big business! So, the Mother of God has to be the one to deliver a message to the ignorant masses. There can't be anyone other than the Mother Mary or Jeshua ben Joseph. And they must look just like the pictures, because if they don't, they are the deceivers you have been told about.

So Mary is very busy in the image business and Jeshua ben Joseph only gets down off the cross to make personal appearances to those people who will carry a truth forward. But he must always keep it in the context of how it has been told over the centuries. He can only change it a little bit. Do you understand the understanding within what I am telling you?

There was a group of people that an angel by the name of Moroni appeared to. He told them to gather up and move out. He instructed them to get away to a place where they could be left alone. In actuality, it was a place that no one wanted. This was done so these people could be seed people. But that too, has gone awry. You don't herd women like cattle and then call yourself illuminated. You don't interpret the word for your own personal gain in that sense. But there is a purposeful good in what they have done. They are prepared for any sort of times that are coming. And at least some of them will survive.

But they are going to be a little disappointed, because in the coming times, the masks are going to come off and people will see exactly what the truth has always been. The images and the expectation of images have to be burned away before truth can look you in the eye.

It's important to burn away images. Otherwise truth becomes very squeezed and filtered, and you only get a little bit of the truth, not all the truth.

These are exciting times you have chosen to live in.

Section Two

Interdimensional Understanding: UFO Making Contact Yucca Valley

An understanding of alien consciousness and the transfiguration of matter.

Consciousness is the all within the all.
Within you, it already is.
You don't have to become God.
You already are.
It's the image you have to burn away
to realize what consciousness is.
Consciousness is eternal and everything that
is the spawn of consciousness is
hooked up to it,
including the aliens.

Chapter 1

Truth and Interdimensional Mind.

Truth is a subjective experience understood through a personal relativity. Everyone has different truth, but every one of you is reading this information to understand something that you feel is a very important part of your life, or something that you deserve to have be a very important part of your life. There have always been caretakers of a truth that creates a continuum. That is called seed truth. Throughout every civilization that has been on this plane there have been individuals who have kept that seed of truth going.

This truth isn't told to everyone, because not everyone is ready to hear it. Do you know you have to earn the right to absorb knowledge? Just because you have this book in your hand doesn't mean you deserve to have this knowledge. You must have expanded the relative base of truth within your own life. No one is keeping this knowledge from you, but you have to earn the right to drink of it. And there are prerequisites to earning this right.

Some of you don't deserve this information because this truth cannot yet be relative to the "box" that you want to put it in, in your own life. You are endeavoring to take it *personally* when it is an *objective truth*. You want it make it personal to glorify yourself and it wasn't meant for that.

You want to make it a religion, a dogma, something that you can be detached from but seek to turn into a religion. And the things that I tell you about have nothing to do with religion. Religion is the pitiful insight into which the human existence endeavors to place those things of mystical nature which are beyond their realm of understanding.

What you really need to have is an interdimensional mind.

Don't you know you have to have an interdimensional mind to know one? You have to learn objectivity rather than subjectivity to have an interdimensional mind. In order to connect to a great mind, you have to *have* a great mind. How do you understand something from a completely alien point of view? (And I do mean alien!) You wish to understand a truth that is a seed that has been handed down for millions of years. You have to be worthy of the seed, and to do that you have to turn the way you perceive things inside out, and make it abstract, not personal. Otherwise, your personal reality will only allow you to see what is subjective to yourself. Therefore you have cut yourself off from a universal knowledge that only a few ever get to know. You have to *earn* the right to learn, and earning has everything to do with having an objective mind when you gain knowledge.

Over this lifetime and thousands of others, you have cultivated an attitude of subjective self. How do you turn that attitude inside out? How do you learn to think as an alien would think? You cannot make contact unless you are worthy of that contact. Otherwise you will simply be a molested victim. There are many of you who have been parlay to an experience that you don't remember because you didn't have the mind capacity to understand. Therefore you were only utilized, used, and never really engaged.

It has taken me ten years on this plane to get people to conceive that God truly does live within you. And even now many of you are reticent to accept that. If it's that outrageous for you to conceive of yourself as divine, how are you going to make a gigantic leap in consciousness to become an abstract thinking entity, so that you can magnetize and equalize to you the same mind? And that's what you're going to have to do. Otherwise you are going to be the spectators of some grand, wonderfully bizarre plan, and you will wind up viewing it as spectators rather than as participants.

The trick is to educate you, so you don't think in the victim/master mode or the personal vendetta mode. You have to learn to turn your own mind around, so you don't exalt yourself, but simply think of yourself as an interdimensional that is capable of interdimensional connection.

The secret is wanting. You have to want to have an interdimensional mind, to the extent that your own personal cause doesn't mean anything anymore. Those *personal* goals of yours must become unimportant.

Encountering an alien must not be another bangle you wear on your neck.

I will tell you something. If you can remember this and apply it simply in your life, no matter where you are in the state of your evolution, you will make gigantic leaps forward in your own progression. Just remember this one statement. You create your reality according to the perception of your consciousness. Your truth is only relative to how you understand truth. And thus, you magnetize to you what reflects your truth, your consciousness. I am not talking about the truth of your philosophy, that doesn't mean anything. I am speaking of the truth that you are capable of living. You attract to you, same. And that is true with every level, from this level all the way to the seventh. It is that way all the way into interdimensional understanding. What you are in consciousness, you will create in your reality. It's as simple as that.

Some of you will only see phenomenon. Some of you will experience phenomenon. Some will truly engage. They can't engage you if you have not created that path in your consciousness. What do you think the path of enlightenment really means? It's not a literal path. It is the *broadening* of your conscious base which allows a highway of reality to connect the unknown to you. You can't get to the unknown until you think unknown.

What you are in consciousness, you will create in your reality. It is as simple as that. And that not only affects your personal life, it also affects the extraordinary and the bizarre. And when I say bizarre, I simply am referring to the unknown.

So let's get this straight. I can't change your reality. I don't have the power to do that. Yes, I am powerful. Yes, I can do a lot of things but I cannot change your *reality*. I can only instigate, and you can participate or not. You are the only one who can say "I want to learn." You are the one who gets filled with knowledge and therefore it is you who gets the reward of the experience. A broadened consciousness exemplifies itself from a limited reality. It becomes unlimited. I can't do that for you. I can't make you have an interdimensional mind. I can give you the information and show you the method, but only you can flip the switch around and be objective.

What will or won't happen to you has nothing to do with words. These are only "words." The people you are desiring to communicate with don't use words. I learned your words to give you the transmittal of a thought, so you could engage a thought and be moved to experience. There is a lot that is communicated to you that has nothing to do with words. These people don't speak as you know speaking. They think. So your learning has to be on focused *thought*. Thought is a language that transcends dimensions. Thought transcends time, distance, space, and every other level. When you learn to "think," then you have interdimensional mind.

Think for a moment. Doesn't it make sense that to connect with all that is known, you have to know? And the way that you know is not what you say, it's what you think.

And if what you think is the absolute truth, then isn't that your pathway through the stars?

What you think. Interdimensional mind. The only thing that crosses all barriers is thought! A thought. A thought is not the word. A word is the aftereffect of a thought. Words are the crippling and limited expression of a thought.

So you must learn how to think. You must learn how to look at the midnight sky and engage a thought with it; that is what pierces through all of the veils.

Thought is the mind's language, and it is objective. Did you know you cannot *think* selfishly? You can only *feel* selfishly. Did you know you can't *think* destructively? You only *feel* destructive. The thought is objective. It's the emotions that make it subjective.

You have to learn how to listen in your heart, and that kind of listening is not about words; it's a knowingness. It's instinct. It's acute awareness. It's sensing the direction of the wind and looking in the direction of the sky. It's *knowing*. It is communication through interdimensional understanding.

You need to spend a lot of moments in absolute silence to bring this about. You have to shut up, shut down, plug up, and just sit, and be aware. Learning to be aware is what will connect you and make a contact. You can try something else. You can build a bonfire. You can form a cross. You can hold hands and chant. You can do all of those things, but it won't do anything. What you want to learn is something you cannot learn on a subjective conscious level. It must come from the objective conscious level. It has to do with your god inside, coming out.

I'm not here to teach you about bizarre beings. I'm here to teach you about advanced civilizations who don't do small talk. They don't appear in the skies just to give you a light show because you are curious. With one blast from one of their ships, these people could send this earth reeling into space. End, over end, over end. So we're not talking about little people who can't wait to embrace you. And we're not talking about people who've come to visit just because you think you've made yourself worthy and you deserve it. We're talking about brilliant beings and the only way you can talk to them is to have an open mind. And the mind that is able to receive symbols and is able to perceive instinctively, is able to pick up signals and is able to see them in the mind. We are talking about beings that have the capacity to talk with you only when you've earned the right to have that happen. Nothing is going to come down unless you have magnetized yourself for that meeting. They have greater things to do. These entities are doing things all of the time.

They are not going to show themselves to you just to prove they exist. They aren't going to give you a ride for $1.50. You have to equalize to you, the capacity to understand. You must equalize to you, the capacity to engage. That's the only reason they would come.

Interdimensional mind is the desire to want to know. An open mind has nothing to do with superstition. It has nothing to do with your fears, your religion, your relics, or where you live. It is simply the capacity to engage the unknown with no expectations as to what the engagement should be, or how it should look, or what you should get out of it.

You would be shocked to discover all the beings that live even within your known universe!

Those that live outside your known universe, you would call ugly, because they don't look like your perception of beauty. Yet their souls are like God complete. They are the life force. We are talking about minds that extend beyond anything you have ever dreamed of. To me, that is God. To you, that is ugly. An open mind has to break up all of that thinking to become interdimensional.

A master who is mastering to Christhood is not simply doing it for this linear life but for all life. You must break down and peel away the limitations and the hysteria of superstition so that your Christ can engage. Somewhere inside, you know you are hooked up to forever, and if you can break away your limitations you can see forever.

Jeshua ben Joseph was an interdimensional being. So was Buddah, Mohammed, me, and a lot more. But that wasn't our aim; it was part of our evolution. You don't get there in your reality if your reality is laced with dogma, superstition, fear and pictures of what things should look like.

You must be done with all of those limitations to engage interdimensional mind.

Everything you think creates,
through energy.
You have within you
the most remarkable machines,
the entireness of consciousness.
You have the power that creates life,
and when that power is drawn up intentionally,
it will create reality.
All you have to do
is give it the pattern you want.

Chapter 2

An Island Called Easter and a Stone Named Henge.

On an island there are 593 giants carved in stone. They are 20 meters tall and they weigh 50 tons each. All of these characters are looking up and out to the sky. Each character represents a human being. Human beings who, in the last stages of their evolution, were able to go beyond the sun. And they knew that this was the last thing they would do on this planet, so they carved in stone forever, a memory of who they were here. They laid down, in living rock, who they had been. You put up a gravestone to mark your dead and fallen. They created a stone in the image of who they were, and that stone says, "Nay, I did not die, but I live!"

The stone was to look to the sky and their testament, their book, was to lay at the base of each stone.

This little island, which is now in the middle of the sea, was once a fragment of a greater land mass. But at the time these stones were made there were no rock quarries on this land mass. There were no timbers. There was only bush, grassland, and meadows. No legion of people lived there, only a handful.

Where did the stone come from? Where did the wood come from? Where did the slave labor come from to erect these giant stones? I ask you to explain that?

On a plateau there are beautiful, precise images of animals and winged creatures. They are laid out so dramatically, it is as if they were stenciled from above. They were. Rumor has it they were created as a religious rite. I say to you, there was no religion. How were these created? How, during that time, was something so precise created?

There are great pyramids on this earth. Some are under water. Some are on land. There is only one left unraped, hidden in the clouds. The stones are precise and well cut. There are no rose quartz quarries nearby. No timbers were nearby. What force would it have taken to have lifted each one to place them upon one another? How was this done so that each of these equates the term pi in mathematics? What sort of civilization could have built this? This planet had advanced cave people at that time. What was the purpose of these monuments?

The North and South poles were once very arrid places. Once there were great mountains where your equator now sits. Buried under the ice caps are temples of magnificent beauty. Their columns are polished, and they are beyond the beauty of anything in ancient Greece or Rome. The paintings that adorn the walls are breathtaking, with inlaid Mother of Pearl and lapis lazuli and fine-streamed gold with no visible mark on it. You would have to look at the extraordinary way they were created and ask yourself, "What barbarians did this?" And under the ice?

There is a circle in a very windy and cold place with no trees nearby. The circle is hinged with stone upon stone, and there is a myth about a wizard in this place. They don't know quite how all of those stones got laid there, set up so precisely. Some people think it was a ritualistic sun clock. Had to have been religion. Why else would they have gone to such trouble? Only God is worth that!

Who built that?

10,000 years ago there was an advanced civilization that lived around the entire world, and that civilization was governed by the most powerful and remarkable teachers this world has ever known. This group of teachers, 13 in all, were endowed with a mind that understood physics, geometry, space, time and distance. They understood God.

These extraordinary 13 teachers that lived 10,000 years ago were seeded on every continent that is now living or under the water. It was their duty to instruct the genetic race of half and half, in areas of education and understanding. It was their duty to lift the species out of mere survival to an understanding of immortality.

They were called the Sons of the Sun.

10,000 years ago these teachers set in motion what is now the residue of civilization.

You have retained less than a breath of what these teachers taught 10,000 years ago. Their sojourn here was 2,000 years long. These teachers never died. They would later be called archangels in myth and legend. They were, in fact, the greatest teachers of celestial origin that ever existed on this earth, and they came to teach, to engage knowledge, before a great cataclysm hit this plane and changed it geographically to what it barely remembles today.

Their students were the seed people of the Ionians, which were later the Greeks, who later inspired the Romans, who inspired the world. Their students also inspired what became the Turks and all of Mongolia. Their students were the original Chinese, the people of Cathay, the oldest dynasties in the world and perhaps the most civilized. These teachers of 10,000 years ago are the ones who taught the students who carved their images in stone, all 593 of them, who set their eyes on the skies and said goodbye to this plane.

All of the 593 people who carved their images in stone are still living this day.

Why would you need such extraordinary teachers? Why aren't they here now?

They are.

What is it in you that insists on destruction? What is that part of your genetic tree of life that insists on destruction of self, the planet, and your capacity for life? These teachers instructed the few who turned that destructiveness over into knowledge and truth, and then left, a long time ago. Shortly after their leaving, this earth of yours went into convulsions that lasted up until 3,500 years ago when your Rocky Mountains were still forming.

What is it that keeps you from being all you can be? What is in your thinking that insists on self-destruction and destruction of this organism that you live on? Because that is what you are facing again.

These people have overseen this speck of dust on the outer part of the Milky Way with its yellow sun, because they have an investment here. The people who were left behind are seeds of those ancient civilizations, and seeds of the teachers of 10,000 years ago. It was, and still is, their job to try to bring the truth to every civilization, and to try to keep you alive when you insist on dying.

They are trying to keep you alive so that one day you can make your own stone and leave this place to become an interdimensional being. But you need to live long enough to go to school to learn how to do that. We are not talking about the school of civilization, we are talking about a school that defies every explanation of civilized truth that you've ever heard!

These beings are holding a lot of things together for you. (They are also called the Constants.) These teachers left you remarkable signs by teaching their people how to do amazing things, like carve out a living stone in their own image, just by thought. They taught them how to move that stone and defy gravity in doing so. These people who carved their images in stone also carved in stone the tablet of their life, for posterity, so that every civilization would remember what they could do.

That island and those monuments survived what happened to this earth. They are still there today. It is the island of giants, an island called Easter.

What do you suppose happened to those tablets they left behind?

The tablets were transcribed onto wood and they lay at the base of each statue. For generations people kept the truth alive by coming and rewriting the tablets, verbatim. For years they faithfully replaced them, leaving them as an epitaph, all 593 of them. Then when religion got to going very strong, and this island was explored in the name of the Church, half of the tablets were burned, because they were deemed evil and inexplicable. The other half ended up in the Vatican. They are down in the cellar to this day.

These tablets told the truth. They told who you are and what your heritage is and what you can look forward to. They dispelled any sort of religion because there was no religion. There is only life, which is God.

These tablets talk about how these individuals created their own image in stone and how they moved them where they wanted them. They describe that where the statues gaze, is where the people went when they left. And the tablets promise they'll be back.

What sort of mind would take truth like that and destroy it? Is it a mind capable of superstition? Yes. Is it a mind capable of fear? Yes. Is it a mind that needs power? Yes. Is it a mind that needs an image? Yes.

These entities you wish to have contact with are the base people of these teachers. And they won't even talk to you if you've got superstition, fear, or limitations in your understanding. Why should they attempt to communicate with you if you can only hear, understand, and believe what fits your image of truth?

The great Stonehenge had living rock tablets that told precisely what it was for. The living rock tablets were burst to pieces during the Dark Ages and thrown into the sea. They were called the work of the devil.

The pyramids held tablets in living rock that told of those who came from beyond the sun to teach about the science of life. The pyramid served many purposes, but perhaps the greatest purpose was that it has survived everything up to this time, even the raping and pillaging of its great capstone and surface. The pyramid would survive everything, in living rock. It is a testament to humanity that the truth is within you.

Pyramid means fire in the center, which means it's in you. It stood as an emblem for all ages to encourage you not to forget who and what you are. It pushes emotional buttons inside you. Its tablets were removed when the Nile was changed and the dam was created. A lot of tunnels were lost; they are now under water. But that will change, because the Nile is going to change again, and so will the dams.

The truth is, the truth was never hidden from you. Civilizations have openly lived here before that were your genetic brothers; that was never hidden. They wanted you to know because they understood how important the preservation of truth is. Today there are only a few places where you can look at something and know its truth, because the ancient books and schools have been destroyed in the name of God, and religion, and power. The books of science and truth and the great library at Alexandria have been destroyed.

Alexandria held these truths and it was burned.

Your planet should have been a cinder 10,000 years ago. It has been cleaned up several times. Civilization has already been destroyed by the releasing of the atom. Did you know there was a civilization that discovered, through alchemy, the atom and its energy sources but didn't know how to harness it? Did you know that had happened before?

It was those green fireballs that restored this atmosphere. There are benevolent beings that have been watching you and caring for you, and they want you to know them. But they can't let you know until you're ready to know. You have to be ready to understand a truth that has nothing to do with religion.

Those stone statues are not religious ceremonial masks. They are the graves of the living. They mark where the people left from.

Those great drawings that are signs that can be seen from above are locations where entities arrived and departed.

It was an airport. That isn't religious! That was an airport. Where did they go? They left before your earth went into upheavals. And they are coming back.

Today there is still a conspiracy that tries to wash out of you the desire to know truth, that essence that is relative. Societal patterning may still be nagging you, saying, "None of this exists." Well, it does. Those teachers are the angels in the books of religion. They live in legend. Their job was to civilize you, to teach you, to give you morality, and a quest to live for. They wanted to make you use your brains when you were, instead, trying to destroy yourselves continuously, because of that little genetic flaw in you that wants to self-destruct. They have tried to turn that around.

These entities would be happy to engage you, but you have to earn the right by removing the mental conspiracy and understanding they are real. They are. And their reality is not for your entertainment. They are extraordinary beings who know what immortal life is all about. They know how to make this immortal life happen in flesh and blood. They know how to engage your mind and uplift you, and they want to do that. I'm here to get you ready and I've been doing that for a very long time. My job is to break down the image in those monkey-mind barriers you've been throwing up around yourselves.

Are they real? Do they exist? Of course they exist. Never show your ignorance by thinking you're the only ones who exist. And never show your ignorance by thinking you're advanced, because you're not.

No one's going to come and take you away in a spaceship to live somewhere else. Some of you have already lived somewhere else. Some of you are part of those genetic people way back in those advanced civilizations that were fortunate enough to learn from one of the 13 great teachers. And some of you are fortunate enough to be back to learn from another great teacher!

I know what you're capable of knowing and seeing, and what you're not. You wear your limitations like clothing. You wear your limited mind like a badge of victimization. I know what you're capable of learning, and there are a few who can make that path all the way through. And there are others who lack the patience and perseverance.

Some of you don't have the time, or the desire, to be humble enough to realize you don't know it all yet, and you won't see or learn anything. But a few of you will, and that demands learning to be abstract.

Those 593 people were born of the womb and of the loins. They were human beings just like you. They were fortunate enough in their evolution to turn the pages of life and keep on turning. And as they did so, they drew to them those great teachers, and that great knowledge allowed them to engage and create, with joy. Do you know they all created their images in great joy? They were going someplace and they knew it. Their next adventure was *out there*, and they magnetized to them the teachers who would engage and inspire their minds to go forward. And they left without dying; they still live this day. They're just like you. Except they had no sense of limitation and no stigma of superstition. Your spiritual dogma can actually keep you here while people are making advances somewhere else.

The hardest thing you have to do is clean up your attitudes. You want to hold on to your death and dying concepts. You want to hold on to your image. You want to hold on to your own stupid ignorance. You want to hold on to what you think the spirit should be. You want to hold on to what you think they should look like and be about. You want to hold on to your little rituals that give you some small sense of importance. Your teachers have had to break through all of that. All of that ignorance has to be torn down and undressed, to rid you of your garbage.

Those 593 beings were part of a very advanced school and they made themselves worthy to go. They wanted to learn, and as a result, the want created a reality. That reality created the experience, that created more want, that created the reality. That reality created more experience, and it just kept growing.

Interdimensional mind is the ability to learn, the ability to gain knowledge. When you can sit out in the black sky in utter peace, and allow the god to come forward, something will connect between you and those great people. Because you'll be pure. Purity comes from burning away the image so you can sit and engage a midnight sky.

The legendary teachers of 10,000 years ago put the seed truth in genetic lines to carry it forward, and they left the seed of truth carved in living stone, so that no matter what the earth went through, that stone would stay there. They knew that someday someone would look one of these giants in the face and be awed, and ask, "How did this get here?" They left those statues behind for the individuals who won't accept the brainwashed notion that they were carved as a ritual to a pagan god.

Yes, you have been taken care of because of that little destructive element in you. Yes, you have indeed been protected, because you have needed it.

Right now your earth is dying. No one wants to make the changes necessary to make a difference. But you have to take a stand! These entities know it and they are herding a few. They are inspiring a few. They are planting seeds of knowledge in some people because they know what's coming. And that seed is a truth that will never be lost in the genetic soul-line of the human being on this level.

And yes, there are some who will leave this plane. They are ready to leave. They may not *want* to leave, but they are *ready* to leave. It is part of their evolution. One day they will simply be gone. No one will know where they went, unless they, too, carve their image and leave it behind somewhere, staring up at the sky. And there are others who are going to gain wisdom, and they will see this earth through to its next step in evolution. Perhaps you are some of those people.

Let us hope, for your sake, that you are capable of somehow turning around your monkey-mind, shutting up, and allowing the thought to come forward that will connect you. And let us hope there are some of you whose consciousness will grow to engage a reality that is bizarre, but most remarkable. And let us hope, for your sake, that you're serious about this. Because if you're not, you're going to be left with your superstition, and your dogma, and your good, and bad, and positive, and negative. You will be left with your little rituals that make you feel better. But you'll still turn on your car engines, and you'll still have things of plastic that you still throw away, and you're going to be destroyed, along with this earth, *unless you change*.

Your brothers are not really trying to take care of a group of kindergartners so much as they are endeavoring to rear up a splendid few. They don't want to take care of you. They want you to get rid of that little thing inside of you that is self-destructive, so that you can get on with evolution. You have a mind that is capable of being interdimensional. They want you to grow up and be what you're supposed to be, illuminated, not in dogma and spiritual mumbo jumbo, but as illuminated beings.

None of the activity of these great beings is religious in its base, although they created the concept of religion. They created the angels that led people to deserts and told them to start new lives. They gave people a morality base. They have done that throughout time, to move people out of pestilence and ignorance and to give you a sense of achieving a moral, basic life. They worked to get early man out of the caves and to stop you from groveling like savages. They worked to get you to walk upright and to give you a purpose. And they gave you the concept of God. And God was in the heavens. They had to do this. Your mind had to have this, in order to advance, but it never had a "religious" base. It was created to make people bloom, and that's all we are doing here.

There is no understanding interdimensional mind. You simply must *be* it. And it happens in the moment of a free thought. And in order to do that, you are going to have to turn your thinking all around. Those beings, those teachers of 10,000 years ago, taught sciences based on thought, because that's where the power is. The power is within the person. And the way to turn around your monkey-mind is to be quiet and sit alone, outdoors, and *be*. And then the thought will come, and that will start connecting you. I can't teach you how to do that, you just have to do it, and be it.

Without external pressure *coming* at you, you aren't going to engage a polarity. Without external pressure coming at you, you are going to have the moment to *think*. That's when you'll connect with these great gods.

These great gods mean business. They won't ever come to you without a purpose. They won't show themselves in the night sky unless you magnetize them, like to like. They won't come without a reason. And the reason would be you moving into their understanding.

They aren't going to get you and put you on a fire and turn you on a spit. They won't eat you. These entities, my people, aren't here to impregnate you either. They've already done that. They are here to engage a few of you. And they make that selection, not you. You have to make yourself worthy by cleaning out the garbage and allowing your mind to be *one*. All you have to do is be able to join in *one* mind. If you've painted your reality full of your pet peeves, your image, your judgement, and your expectations, you don't have room for them, and you can't possibly have an experience.

Bliss is having the grace
to discover a moment that is without
the illusion of the past or the future.
It is a moment, collapsed in time,
to be that moment.

Chapter 3

The Art of Being. One Simple Way to Overcome Monkey-Mind and Just Be.

The process of overcoming your image and your need to judge is not easy for you. A simple way to begin this process is to move outdoors to a place where there are no other people to distract you.

Sit down where you have no worry about who is in front of you or who is watching you from behind. If you must go with other people to feel comfortable doing this, sit in a line, not in a huddled group. Take whatever you need to feel comfortable to this little outdoor retreat. Take something to sit on and something to lie down on.

Sit there, in quiet, and still the mind until nothing exists but you and eternity.

After an hour or so of moments, after you've gotten into the swing of things and begun to conceive of just being, a moment of knowing will come. In that knowingness you will feel a lifting of your being that comes from knowing.

This is very intentional. You need to know that this will happen. In the knowing, you will pick up a sense. You may get geese all over your body. You won't know why, but it is because you are clearing out some of your stuff. You must stay there long enough to get over all your thoughts.

"Why am I sitting here?"

"I thought I'd see a spaceship by now."

"Aren't they going to take us for a ride?"

You have to sit there long enough to get over that cycle of thinking, because only then will you have the thought. When you have the thought, the understanding, look up immediately, and you will see a wonderful sign in the sky. It is a beginning, that is all it is.

This is the beginning of the process of "being." Being is simply allowing you to be removed from the subjective mind of the image. *Being* is one of the greatest teachings there is, because there is an art to being. Being is simply *allowing* you. It is allowing yourself to be removed from the subjective mind of the image. It is just being.

Being, is being one with the night wind. You can feel it. You are it. Being is when you look up at eternity and there is no separation between you and eternity. That's being.

And there's nothing that says, "Let's talk about what we just saw." Just be, don't talk about it. You need to just "be" a lot. Because in just being, you will learn what having an extraordinary mind is all about. You have to experience it.

Sit as long as you feel like sitting in this place. If you want to go to the bathroom, and you get hungry, and you want to eat, and you get tired, and you really want to go to bed—get up and go do it. Don't sit there just to try to prove to a friend that you really want to have this knowledge. Face it, if you just want to go to bed, it's all right. It's better to be honest with self than to keep sitting on a hill in the dark because you think you have to. That will get you nothing except a victimizing reality. But you should stay there as long as you can.

And remember, the greatest activity happens when your night shifts to morning.

In my day, we had more stars visible in the night sky, and we had two moons. Well, one was actually a ship. It just looked like a moon. But there were two spheres in the sky.

Not everything in the night sky is what it seems to be.

Never show such ignorance as to think that out of ten billion suns in the Milky Way, that this pitiful little speck of dust in the outermost corner of God's mind is the only place that knows life.

In my lifetime, I sat outside, on a very large rock, and I contemplated the enchantress moon. And I contemplated another moon. I contemplated it quite a bit, except its journey was quite irregular, not like the enchantress. I sat out and observed and watched. That is how I learned.

Your image and sense of social
consciousness only knows what the
conspiracy of civilization has told you
and no more.
You have literally been taught to disbelieve
anything other than that.
You have "learned" to destroy anything
other than the way society wants you
to view God.

Chapter 4

The Computer God. And a Tale of Two Cities.

In 1956, there were some explorers who airlifted supplies into New Guinea. One of their crates was seized by a very secret tribe of people.

They are not a civilized people, like you are.

The tribe opened the crate and found all sorts of wonderful things they'd never seen in their lives. There were tins which contained food stuffs and tins with tobacco and instruments. And the most prized possession in this booty was a map.

It was a road map with New Guinea on the front and the hemisphere on the back. This tribe was so uplifted by this acquisition that a whole new cult raged throughout all the tribal sects. This was big business that had happened.

They had never seen a cargo supply box. They had never seen a tin with foodstuffs in it, and they had never seen instruments or tobacco stuffs. And they had certainly never seen a map, and they had never seen a silver ship drop such a treasure from the sky before. A whole new cult began to rage, and they began to dress, as best as they could, like the pictures on the tins. And they used the instruments in sacred rites and ceremonies to the gods. These navigation instruments and road building instruments soon were considered sacred ceremonial artifacts. They considered the map to be the code from the gods telling the natives how to find God.

When the explorers went to receive their cargo, they were received as the gods from above. They were given the tribe's best women, their best goats, and their most precious metals. And the natives fell down and began to worship them. They even situated these explorers in the temple they had built around the artifacts.

These were, of course, uncivilized people.

One of the items on board was an airline ticket. This particular ticket had instructions about how the computer had picked seats. This language was translated to the natives by the visitors, and it became their mantra, because they determined that the computer was the supreme god, because the computer had chosen destiny.

We sigh and pick our teeth and think, "Well, that's because they didn't know any better." I tell you, you are much the same way. You look at artifacts left from ancient times and you want to connect them to deity, to personal destiny, to God. You are, in essence, the same as this tribe, who found a few things that fell from the sky and began to take them personally, believing that their whole destiny was determined by a computer in some far city.

When I told you that in order to have interdimensional mind, you had to learn not to take things personally—not subjectively, but objectively—I was telling you what you needed to know. These entities are not here for a subjective mission; they are here for an objective purpose. Their purpose is to balance out this corner of the universe, which is currently inhabited by wild, uncivilized people that are not nearly as sweet, and simple, and calm, and delightfully ignorant as these tribespeople, but who are, instead, deadly, insisting on the destruction of everything.

If you desire contact for personal glorification, or for purposes of worship, then you have separated yourself, in consciousness, from great mind. The need for personal glorification separates you from great mind. Everything that is seen *outside* of you separates you from the ability to engage them.

Consciousness must be objective, in as much as it must have the ability to be abstract. You must have the ability to observe, and be one with what you observe, not be less than, not looking for a savior or redemption, but rather looking for an equalization, an opportunity to move forward in dynamic knowledge.

Where are the remnants of past great civilizations that would prove these teachers existed and interdimensional contact had previously occurred? The remnants are not easily identified as machines as you understand machines today, or as "learning" as you identify learning today.

A German scientist recognized that flattened stones he saw in a museum in Baghdad were not religious relics, but were ancient batteries over 2,000 years old. Had he not realized what they were, they would still be sitting in a museum, identified as ancient religious stones.

A discoverer found ancient maps in a cave that dated to the first century A.D. The maps clearly indicated the area of the Mediterranean that was subjective to that person, because there was no "known world," and yet the map was strange, because it had the Mediterranean, the Arctic and Antarctic, both upper and lower Americas, yet it was an aerial map. There was another one found about the time that Columbus sailed for the unknown. (If Columbus had found this map, things would have been a lot easier for him.) This map was tucked away in a cave, too.

In the Yucatan, in Bolhezia, there are caves that contain paintings of all the stars in the system. They were done over 15,000 years ago. There are a lot more stars in these paintings than you see now, and yet on them are inscriptions with lines and coordinates that point from Venus to this earth.

Who did those? Why are they in a cave? Why aren't they out where everyone can see them? Because caves are an obvious natural hiding place for truth, a place where sooner or later the curious human being will stumble. Besides, everything that was built openly as a truth for you, you have literally destroyed, burned, or washed away. You claimed they were demonic, satanic, or paganistic.

Where did the gods come from? They came from several places. The gods are merely what you are, advanced. They are exhibiting the highest psyche of the human formed in consciousness. They are beyond the barriers and limitations of subjective, petty beliefs. They are pure knowledge in the grandest sense of the word.

According to the history of the ancient civilization of the Eskimos, the first civilization of human beings was the greatest civilization that ever lived because they were human beings who possessed all knowledge. They clearly understood the nebulae and the four corners of the firmament, and they had destined themselves to go back and forth in space. This civilization was based here on earth. The Eskimos are not originally from the frozen North; they are originally from Ceylon. How did they get where they are? Part of their people were deported from that base, and their ancient history speaks of great teachers who came in silver birds and transported them to the North. And to this day, they wait for those teachers to return.

There was a great city near a lake that is now high in a plateau. There were two great cities that existed there, and their people were very advanced, quite like you. They had the same flesh and blood, the same eyeballs; they all went to the bathroom the same way you do. But they were very advanced. Except they became so advanced they created a very decadent religion, one that was subjective to human misery and pain, and the separation of human being from human being. They had discovered, through alchemy, the way to disperse matter and release its energy. They had the power of the bomb and they used it.

There was a great decision made in the consortium, the kind of decision being made now. The decision was made to do away with that genetic strain, because those people were breeding a collapsing consciousness in knowledge. They were breeding a decadence of the human being. They were dying in consciousness, and worshipping knowledge, only to the degree that it served them as powerful people who could subject others. And they were killing the life force on the planet.

There was one mortal allowed to ride the ship that day to record what happened. And what was seen is recorded in my people's ancient books, *The Ramayana*. One man rode this great and splendid ship and watched as, in one terrible moment, a blast came as a column of white fire, more powerful than ten thousand suns, and in the twinkling of an eye, it wiped out the two great civilizations that had begun their decline on this plane. They were gone in the twinkling of an eye. And one man rode the ship and watched it happen.

The survivors' hair and fingernails fell out, and their flesh fell off of their bodies. Their animals and birds turned white in color, and their water was poisoned, and so were their fields. In a moment, their genetic line was gone forever.

Cruel? I want you to wake up and know what you are dealing with here. Was that cruel? What about free will? Was this God enacting pain upon someone else? Who judges this?

That plain still exists by that great lake. There are caves there that show remnants of what this civilization looked like.

The decision was made because these people had released the energy of matter. They had clearly fallen into decadence. They had fallen into the worship of the physical body, and into what you call power. They ruled their part of the hemisphere with atomic energy. But they were decadent people who were ruining and threatening the whole of the world.

It's that little chromosome; it's that self-destructive chromosome that they had. And they would not advance any further than what they had gained in power.

The decision was made, for the good of the whole world and for all civilizations to come, that this, the nemesis of civilization, should be destroyed. And in a moment, they were gone.

And this wonderful man who rode the ship, wrote about it in his text.

Where did religion come from? Where did this picture of a terrible God that would punish you if you weren't good, come from? This is an example of the thread that has lead you to that conclusion.

What I have told you is not a legend. It is the truth. This is not a myth. What happened to these cities did not happen because of some ridiculous dogma. It did not happen under the auspices of some priest or any law that separates human beings. It did not happen because of any law that creates dissention and war and diversity.

The teachers have always said, "I will inspire and teach you, and uplift you beyond your flesh, to give to you the waters of eternal life, to give you the knowledge that will allow you to understand all of the mansions that are in heaven. But in order to achieve that, you must stop being a primitive, and start being the god that you are. If you want the knowledge, grow up, and earn the right to learn it. And furthermore, you must live it. It must be a living fire within you."

That was the law.

And these same beings, these teachers, have overseen your civilization through all of your known history. They have kept you from destroying yourselves or from falling into complete decadence. Do you know that when the mind stops learning and the spirit is no longer engaged, the physical being is run by the hormones? It becomes decadent because that becomes its reality.

Sodom and Gomorrah were destroyed by an atomic blast from a great ship in the twinkling of an eye. Why? How low does the human spirit have to go to engage sexual intercourse with the cattle of the field as a religious rite?

You are so endowed to be divine that you destroy your own divinity. You are so endowed as divine beings that you destroy and corrupt your own way to learn knowledge. You cut off your lifeline for survival; you destroy the very earth that is your home. So great are you, that all that is only relative to your own universe happens in your consciousness, therefore your life. Ponder that. You are so powerful that what you create in consciousness, is what you create in your life!

All these great beings have ever done is to try to keep you from hurting yourselves and collapsing on yourselves. Heard of divine intervention? I've played a big part in that number.

And stupid people cry out, "Free will! What about free will? How can you intervene on free will?"

By grace, to save you from yourself.

These beings are not to be worshipped and they are not to be idolized. They are the gods of Moses' day and they are the gods of Buddha. They were the gods even in the time prior to Buddha Amin. They were the gods of 10,000 years ago. They are the gods who set truth in motion. Are they suppressing and keeping the door to eternity away from you? No. They are endeavoring to engender in you the ability to wake up and make that connection for yourself, so that you can bridge the light. To create the bridge of light is to create it in your consciousness, that's all.

Does it bother them that they could destroy one city? No, it doesn't. Why? Because their mind is capable of understanding the foreverness of spirit and thought. They are not petty or small-minded enough to think this is the only life you will ever live. They could destroy you genetically, only to raise you up in a greater physical form that the spirit can engage. That's all they have ever done.

I am trying to explain why and how you must move your mind from the subjective to the objective, and that is very arduous. You must learn to become one, and go forward.

Why do you think you were born at this time? Do you really think it a coincidence that you were born during this time with so much happening? Come now! There is a voice within saying, "Let us strive to be a better person. Let us push to have a greater mind." Push your capacity for greater thought beyond dogma and petty rituals and become brilliant. It only takes a moment.

Does it take being the only person on a hill watching a whole city go up in flames to start you thinking? Is that what it takes? Or can you start to think in a quiet moment under a midnight sky? Can you understand that all your petty beliefs are illusions and that you are connected to the same divine mind that all these teachers are hooked up to as well? Can you comprehend that the pathway to contact is the great god that all of you are? It only takes a moment to realize that, and when you really embrace it, your whole reality changes. That's all these teachers ever endeavored to inspire.

One day these teachers packed up their divine belongings. They took all of their great knowledge and left, hoping that the seeds of truth would begin to flourish in the garden of humanity. Because you, after all, are their brothers and sisters. No matter the difference in flesh, you are, by God, brothers and sisters to these people. They have been waiting for this seed to germinate and grow.

During the Dark Ages, 250 million people died in holy wars. And every ancient school, and every ancient text of truth, in stone, papyrus, or leather, that could be found was burned as demonic. Truth was destroyed. That is why that period is called the Dark Ages.

People within the Dark Ages were illiterate. They could not read or write. And yet in the days when the Greek Empire flourished, all people were literate. And yet the Greeks worshipped gods, not the one god. The one god many of you have been worshipping wants you illiterate. That is the interpretation of those who rule religion. They want to keep you suppressed, so that truth cannot flow out of you like a river. And during the Dark Ages, all that could possibly be destroyed was destroyed.

Why can't you find any of these relics now? Because of that little need to do away with anything that would intimidate you, and make you feel less than.

Your scientists today have one unspoken, but well accepted rule. Never ever do a paper on anything that is outside the realm of acceptable fact. That is their unwritten law. Even your "factual" world is guarded, to keep the unfactual out of it, to keep you sterile and under the yoke of work and labor for those who rule this world.

What do you think these teachers think about this? What do you think they think when they look upon the fishes of the sea, washing upon the shore, rotting with lesions and ulcers? What do you think they think of great green forests that stand naked and barren and dying? What do you think that they think of the cities, where it is a normal consciousness to destroy human life through drugs and strong drink and sexual perversion? What do you think they think about cities where stealing is a normal way of life? And when I speak of theft, I refer to everything from petty theft on the street, to the theft in the highest places sanctified by law. What do you think they think of how you're living? What do you think they're seeing when they look at your civilization? Do they see utopia? Do you really think they think you're swell? No.

At the present time you have the capacity to make a 500 megaton cobalt bomb. That bomb can destroy the entire world and every living organism on it. What do you think the great beings think about the people that own that secret? What is the consciousness of the people who hold these secrets? Would you trust your neighbor with a cobalt bomb? How about your children? Would you trust your mayor with one? How about your worst best friend? How many times have you gotten angry because your little image was touted? How many times have you gotten furious because your altered ego was bruised? How many times have you been so angry because your altered ego got bruised that you wanted to violently strike out and hurt someone? How many times have you felt like raising your hand and doing away with everything, to make someone feel sorry for what they did to you? That essence about you is what is so dangerous. And it is that essence that keeps the door locked on you and eternity. Just imagine that little essence having a cobalt bomb.

Do you think I am proliferating fear in the marketplace? I am telling you the truth.

Why is it necessary for you to read this now, at this place in your life? Because with knowledge, you gain a greater insight, and that greater insight really is your salvation. Knowledge is what is going to save your hide. Truth will always be preserved. And if it is preserved within your god, and your god is manifesting it, and living it, you will be preserved. It is as simple as that.

And to understand that, you'd have to see who is working with you. The great entities do not want to see the fish in the sea, whom they love, destroyed by your need for plastics and sewage and garbage.

Think about this. I know you think my people were barbaric, but I didn't pollute in my time. I had an army that was 2 million strong, and our dung and urine went into the earth, and was recycled. We didn't have heaps of it, laced with chemicals to break it down, only to then throw it into the sea. We didn't have plastic and papers. We didn't have the conveniences of auto machines. We had two feet and great steeds. And our ancestors had great ships of light. We did not leave a trail of pollution.

What we ate came from the earth and went back to the earth.

Knowledge isn't dependent upon the number of conveniences you possess. That just says you're lazy. Knowledge is based on the concept of the awareness that you have. Knowledge gives you the truth that makes you aware of the conditions of what you are. You are reading this to open your mind and help get you beyond your silly little image. And the image is the very thing that caused civilizations to perish.

Simple people were always selected by these teachers, not because they were simpletons or because they lacked brains. They had the brains, but they were simple enough not to cloud truth. They had no "image." The truth can come through a simple person purely. Simple people were selected as prophets, workers, and as the people through whom the truth was spread.

Knowledge isn't about going to school and learning a theory that is, at best, an hypothesis. Knowledge isn't about how many crystals you can wear or how much esoterica you can memorize. Do you know what esoterism is? That "ism" means it is a truth, an observation, a parting of the waters that is taken as a personal act to reveal something sacred within the individual who recorded it. That is esoterism; taking truth and making it personal. But if you turn it around, and release the personal off the truth, then you will have the knowledge that will connect you to the stars. Because your personal need—personal, personal, personal—to think things are done for you, inhibits your mind from growing in consciousness to consume that mind that is "out there."

Some of you have been studying about the image and some of you have learned to break through and broaden consciousness. Those of you who have learned to do that can validate what I am about to tell you. Dispelling the personal is what connects you to unlimited mind. These entities are going to make contact, but they will not make contact with an esoteric person. They will not make contact with a person involved with the cosmological side of spiritual truth. They will not make contact with an entity who has adorned himself with the garb he thinks he should be received in. Remember the tribe that dressed up and created a place of worship for the great ship that dropped the cargo out of the sky? Do you think they're coming here to take you back with them because you dressed to look like them? You couldn't dress to look like them; you have no idea what they look like!

They are going to make contact with a simple mind, and yet it is that simple mind that is the fire of genius. You cannot illuminate a philosopher. You cannot illuminate a religious fanatic. You cannot illuminate a seeker, because seekers always seek. If you illuminate them, they stop seeking, and their thing is to seek. You cannot illuminate an image, that little thing that you tout that says you are you. The image is what makes you conceive of love and hate and like and dislike and beauty and ugly. You cannot illuminate an image; an image is what it is. But a simple mind, you can set on fire.

These entities are going to be even more visible in the days that are coming. They are going to become even more blatant. They are looking for the seeds, another step in preservation. And a prerequisite to being a seed is having a mind capable of crossing all barriers in understanding. They are not looking for minds that say, "If you can't say the word then I can't possibly know it." They are looking for a mind that gets the *impression*, the knowingness, the instinct. They are looking for the intuitive, the mystical. They are looking for mystical minds.

The greatest civilization that ever lived was based first on the mystical mind and then on scientific fact. Because it is only the mystical mind that opens up the doorways. Their contacts were their partners. That partnership is happening for some people today. They give to you, share with you, engage you, and you give back. That is what it is, and there is nothing religious about that. There is nothing religious about an airfield. There is nothing spiritual about an airfield. Spirituality is in the quantum nature of reality that is mystical mind.

These entities are becoming even more obvious, and no one can excuse them any longer. Then, of course, there will always be that sleepy, hypnotic part of society that will buffoon anything. No matter if they're looking right at it, it doesn't exist. Because they've been trained to get old and die, and they have been trained to only think one way, in case anyone in high places needs to set them on fire, to inflame them, to move against whatever enemy they want them to move against. Whatever the government wants, middle America will move against in the name of religion. That's the sword. Get middle America angry at anyone, and soldiers will march, in legion, over innocent people.

These entities who are showing themselves in the midnight sky are here to take the seeds, and to start working with knowledge.

The process of consciousness and energy that I teach burns the image and allows the consciousness to float in the now. That consciousness has no past and no future; it is a *now* feeling. It is an awareness, and it is in that great awareness that you start connecting to great mind. It is as if you turn your lights on, so they can see you. And they will.

Being is working consciousness to the abstract, and in order to do that, you have to burn away the image to get there. Because it is only in the abstract of consciousness that you hook up to interdimensional mind. You can't hook it up visualizing a dream. You can't hook it up visualizing a want. You hook up when you're open.

The simpler the mind, meaning the ability to expand consciousness, the greater the ability to have a partner in life. Do you think this is a fantastic truth? Let me tell you something. You have had civilizations that have disappeared in the twinkling of an eye because of this profound knowledge. You have had civilizations that are genetically seeded with these people. They are trying to breed in life and breed out that destructive, ugly, touting, little need for the image, that destructive force. Some of you are human beings whose genetic tree is not a linear process on this earth. But instead, you come from beyond the sun.

Could it be your god has selected this genetic line in order to hook up?

If you have followed your inner knowing because it felt right inside, even when the world laughed at you and questioned you, then you are going to hook up. Because the god within understands what the image and social consciousness could never know. And that's because your image and sense of social consciousness knows only what the conspiracy of civilization has told you and no more. You have literally been taught to disbelieve anything other than that. You have "learned" to destroy anything other than the way society wants you to view God.

Your leaders would tell you, "We have to stamp out evil." What they are really saying is, "We have to stamp out truth."

To get to that great place in your *mind* demands clearing consciousness, and you don't clear consciousness by fasting. And you don't clear consciousness by abstaining from life. And you don't clear consciousness by visualization. Visualization is only the mind creating another illusion!

Don't you understand how great your mind is? You can create light in a room and yet it can be an illusion. And yet that doesn't do anything because you are unchanged. In connecting with these people, you are being objective, non-personal. That is a consciousness that does not intend to glorify, but to engage. And that's what you need to magnetize the great ones to you.

You can't begin to know how they think. Because you can't imagine doing away with two civilizations in the twinkling of an eye, could you? Do you know why? Because in reality, you still think life begins at birth and ends in death. In reality, you don't understand the immortality of the spirit, and yet you call yourselves spiritual.

If you don't understand that, then you also won't understand why nature would violently destroy everything in its path to make a change.

I will tell you this. There have been things in nature that have occurred that have destroyed whole civilizations in four hours. What would you think about the earth if it rotated on its axis and, in a moment, sent all of the oceans blasting over all of the land masses, killing everything in sight, including human beings who are supposed to be divine? And what do you think about an earthquake that kills people? Does that mean anything to you? You don't pay much attention, not really. Not when the people weren't kin to you. Oh, you feel a little sad, but tomorrow there will be something else going on. You don't remember the people who perish in earthquakes because it isn't personal until it touches you directly. If it's personal, it's a different story. What would you think about nature if it had the same attitude? If you were nature, could you take that sort of action?

Let me give you something to think about. When you get a sore on your leg, you scrub that sucker, and you dab that sucker, and you bandage that sore and you wrap it. Because you have to kill the bacteria. And yet bacteria has a consciousness. What is your reason for the action you take? You don't want your wound to get infected. What do you think will infect it? Another life force! But you will destroy bacteria because you don't want it to destroy you.

Answer this question for me. Are there super lives and low lives? Is there important life and unimportant life? Isn't the life force, the life force? Isn't it all the same thing? Yes, it is. You cannot be eternal, and be in the *now*, and have an aware consciousness, and not know that the life force simply *is*. It is as powerful in bacterium as it is in you.

In great mind, there is no greater or less. In great mind, there is no separation.

You couldn't destroy a decadent society, but you could wipe out entire civilizations of bacteria with one stroke of soap. You don't care. You don't know them. It is an impersonal act. But you do it in an effort to continue yourselves. And somehow, you know that bacteria will come again, and breed somewhere else. That's the way it is, isn't it?

I wonder what I'll be accused of doing giving this teaching. I can hear it now. "That Ram, he has no consciousness! He's creating fear in the marketplace. He's talking about destroying whole civilizations and equating it to bacteria!"

Well, it's the truth. And the truth will set you free.

Don't be a hypocrite! If you can do away with the bacteria on a sore, you can destroy civilization! You must understand, the life force is a continuum. Entities that perish are spirit in a moment, and they are caught back up into the light. And who are you to say that an entity from one of those destroyed ancient civilizations didn't evolve in that experience and come back to be you today?

Afraid of atomic energy? Did something go off inside of you when I spoke of ancient civilizations who had that power? Are you afraid you could have been one of those people? Yes, you could have been. Don't you understand? The life force is simply recycling itself. And you, spirit and soul, are it.

How heavy are you in body? When you think about dying, it will tell you how heavy you are. How great are you in spirit? When you think about living, that tells you how great you are.

Are you ready to connect with an abstract mind that doesn't feel bad about wiping out ten million bacteria? Does that put it in perspective for you? For all of you doubters, who's to say those bacteria didn't have wife and children? And who would say they didn't leave families? Who's to say they weren't pregnant? More than likely they were. You kill unborn bacteria! Do you understand?

I know this is stretching, and I know I'm truly treading on thin ice here, because your image wants to turn this around and see this as information without love and conscience, because this is scary and unbelievable. Well, hypocrite, the next time you have a sore (and if you're a hypocrite you'll get lots of sores) and someone tells you to get rid of the bacteria, remember this teaching, and the knowledge about what the life force is. Because everything is God. Isn't that the premise of understanding all there is? So if everything is God, what isn't? Who would determine what isn't God? Are those great teachers not God? Yes, they are. Are you? Yes, you are!

Let me tell you one thing about what you've just considered. If you contemplate what I've just suggested, you would win one small victory over death. Your own. Because in reality, when you change consciousness, you change reality. And if your reality is coming from your consciousness, which it is, and you understand that a whole civilization can die in the twinkling of an eye, but still live forever and ever, then you will never die, because that will become your reality.

Ignorance. Do you know how many people claim to tout the truth but it's just recycled ignorance? Good and bad and higher and lower and separation; God out there somewhere, instead of God inside. People spend their time meditating with their hands open, trying to receive something. What an hypocrisy. God isn't *out there*; he is *inside*. God isn't outside of you. God is inside you. You live as if your power were wrapped up in some inanimate object. It is inside of you.

How ignorant have you been? How closed off is your mind? Is your whole world your image? If it is, you might as well bag it, and leave this place, because you will never learn any more than that. You aren't capable of it. You have plugged yourself up with taboos and superstitions, with ignorance and prejudice, and you have no room left for illumination. Because you are not simple; you are complex. And that's no longer a compliment! Calling you complex shows your ignorance.

Listen, this is not a matter of going out and engaging the stars because they haven't already engaged you. It is a matter of learning the knowledge that gets you to inter-dimensional mind by opening up the mind you have. These great beings are not right or wrong for what they do. They are overlords of a very feeble genetic strain that seems intent on killing itself for greed. That is it. They want to raise your consciousness and locate those who can carry forward the truth.

They aren't going to show themselves to you because you own lots of condos and bonds and securities, or because you consider yourself powerful. When they have a ship that can spin this earth out of orbit, do you think they'll be impressed with your power? No. Do you think they'll be impressed by your virtue? No. Because what do you base virtue on? Are you a hypocrite? Hypocrisy is not virtue; it's ignorance. And that's not illumination; that's a stagnated mind. They are not going to waste their time on you without a purpose.

They will show themselves to those who will be the seeds of truth and who will later build a bridge of light that will appear and disappear to create the communications.

You're not here on this plane to save the world. You're given that opportunity every day of your life. (This will hurt, but you need to hear it.) Every time you turn on your automobile, you are killing the earth. Every time you purchase things in plastic, you are killing the earth. Every time you flush your toilets, you are killing the earth. Every time you throw away your disposables, you are killing the earth.

"But what am I to do?" you ask. "That is the way it is."

Change it! Because every day of your life, you have the opportunity to change it.

They are here to try to keep procreation going. Every day you say you're all for that, but you don't do anything to evoke change in personal self. So you are a hypocrite!

That stung, I know it did. But you must hear the truth.

We are talking about the mind that is intent on raising you up, a mind that is intent on engaging you, in terms of knowledge. We are talking about entities who blasted away civilizations 10,000 years ago. We are talking about entities who left the truth behind, only for human beings to burn it.

Change. Being simple is simply reflected in your reality. It's not enough to say these things, you must *be* them. And if you're having to contemplate them, they're not in your consciousness. You are forcing yourself to do something you don't want to do. That's not evolution, that's stifling self. If you're going to be a hypocrite, be one. And tell the world you're one. If you don't want to change, don't change. You're respected for that. It doesn't affect the portion of love meted out to you. But to get to where you want to go, it demands more than just sitting on a rock and waiting for a blast of light.

Consciousness is not linear; it is *all*. You are not loved depending on how much you love God. That is an asinine concept. You cannot love God unless you love yourself. And if you love yourself, then you are connected with the One that is the *All* in the mind. And when that is the case, you change, and your life becomes a personal fire.

I am very passionate about this, because I see your beauty. I see the gods that you are. But I also see that nasty little overlay, that image that doesn't care. I also see the civilized, industrialized persons. When will you burn that away and take your own stand? Because that's what it takes.

If these entities had not intervened on this planet throughout your civilizations, it would be sterile today. Did you know the planet Mars was once like this earth?

What you have on Mars now is an example of what happens when you leave human beings alone. If you haven't seen a good photo of that planet as it is today, you should go find one and look at it.

There is a real intent to keep this planet going for the people who are here. You are going to see spectacular things, and some of you will be fortunate enough to be a part of them. But you are living in incredulously dangerous times, much more dangerous than you can imagine.

This mind of yours that is reaching out into the great unknown must also *be* that great unknown. It cannot be separate from it. The truth is, if you love your earth, then don't buy a burger in styrofoam. Buying a burger in styrofoam is being separate from truth. Being one with truth is buying a burger and having it placed on your hand.

Look at your hands. They're washable. You never have to throw them away. They come with you wherever you are. You don't have to pack anything special. You can just walk up and say, "Put it in my hand." Isn't that wonderful? Do it!

That's a small example, but it is an example of not being separate from the truth. It is an example of being one in consciousness, so that the word is a living word. That's what they're after. Simplicity.

Do you know what a complicated person would do to this? They would take this truth and say, "Well, theoretically that would certainly be the thing to do, but economically speaking, and according to our trade deficit, it would not be wise to altogether stop using these disposable items. After all, where are we getting our meat from? South America. And after all, is that not a national product we should engender, to help them pay off their debt to us? And if we really do what you're suggesting, we would have economic collapse and the bond market would fall." And the bond market owns 80% of America, did you know? Now, that's a complex way of thinking. (Those are the kind of people you just want to shake.) The simple truth is, there's no complexity. There is just the Is.

"Just put it here. I could take the paper. The paper would go back into the earth. But why don't you just put it in my hand, because I can wash this. And when I leave this earth, it's biodegradable. One day my bones will push up a flower by fertilizing it."

Now, I know you're wondering what this has to do with interdimensional mind and the entities you wish to contact. Believe me, it has everything to do with contact. Prophets rise up from simple people. The simple people are the seeds that get preserved, because in them, truth is the light that lives on. It is through simplicity that genius is born.

Complex people don't have room for genius. It's not economically feasible in their time flow, in any one given day in the business week. Complex people, smart people, are not geniuses; simple people are. The Industrial Revolution came out of simple minds who were later touted as geniuses.

If it's hard for you to hear about the destruction of Sodom and Gomorrah and Antruschon and Eloman, then you aren't ready to hear everything there is to know. And that means you do not have the mind capacity to go further than that, because you are emotionally blocked. And that's all right. You will learn what you are capable of learning, and you will see what you are capable of seeing.

Did you think about the life force when you contemplated the destruction of bacteria? Did you think about it being equal to the destruction of a whole civilization?

As long as you have a rigid, dogmatic truth that you feel is in the name of love—which it is, but it is an ignorant love—you are not going to go very far in consciousness. Because if you have to look at that, it becomes a great block in your way. How can you engage the unknown when you go into the unknown with rules and regulations? When you go into consciousness with rules and regulations, all you get is support for those rules and regulations.

These great beings are not here to put you on a
spit and turn you.
They don't eat you.
The Great Ones, my people,
are not here to impregnate you.
They've already done that.
They are here to engage a few of you.

Chapter 5

The Ships. The Beings. The Relativity of Free Will. And God.

The Ships.

The great ships are made of several different metals, most of them foreign to your earth. The metals are light; you could hold a part of one of their ships on your shoulders. They are conductors of force fields that are created at zero degrees in a vacuum. The vacuum is created by a rotating motor, which is a perpetual motion engine at a vacuum of zero degrees.

They travel on a pathway of light. The auric field around the ship is determined by the degree of speed with which they move. There are ships of every description and nature. In ancient times they were termed chariots of fire, and they took selected beings into the firmament and never came back. They have been seen as clouds, and they are; and they have been called pillars of fire, which they are. They come very small, one person style, and they come large enough to cover half the sky in any given night.

Their crafts, though luminous and wonderful, are also dangerous. They emit, at a certain frequency, what we will call RPM, a very dangerous field you call radiation. It is at this RPM that they are the most stable and they oftentimes burn the earth, eliminating growth and vegetation in that place for many years in your time and counting.

At the highest frequency, you can be on board and be unaffected, yet a human being could not tolerate the degree and speed at which these vehicles travel. Therefore human beings must be impacted with a substance that is like jello. Beings are fed this until it fills all of their organs. They are then placed in air-tight containers and fed oxygen. It is the only way you can survive an interstellar trip. They are, indeed, capable of taking you places, but not to your harm, especially if the being is in partnership with them.

They love, but to a degree that is different from your understanding of love. They love the life force, knowledge, truth. Because they *are* it.

They are not lovers of an individual person, by degree. They are lovers of the whole. That enables them to have a mind that is fully opened and bloomed. It allows them to integrate.

These beings and the capacity of their ship varies according to their mission. If they are here to extrapolate your genetic pool, then their ships are designed in a very bare, essential nature. If they are here to engage geographical, or geological testing, then they have much larger ships, equipped very differently.

They often hover over bodies of water for the ionization of the water and often over power plants for the leakage of power that comes from them. These people will not wave at you or stop and say hello. If you happen to watch their great ships zoom on by, you can assume they are on a mission somewhere. And if your eyes can follow them, then bully for you.

They have the ability, through raising the vibration of the vacuum, to become visible and invisible. They can appear in the middle of your sky, go for a short distance, and then disappear. They have what is termed "cloaking devices," but it's not really that. It is simply a part of their apparatus. At a certain speed, they do disappear because their vibration is no longer in a three-dimensional understanding. They can go to the other side of the Pleiadian system in the counting of five, and they disappear to get there in that time. They are beyond light, distance and time. They can go beyond the sun in a moment. They can go to the parallel universe through the black holes. They have been everywhere. They are the explorers of the all-in-all.

The ship of my people is a great ship. It has smaller ships that are like handmaidens. The ship is invisible at night, except for the bottom lamps. It's bottom lamps are often thought to be separate stars or separate craft. It is made of metal that is the color of the midnight sky. When it is illuminated, it is brighter than ten thousand suns. The only insignia that it bears is the triad.

It has beautiful lights that change intensity according to its movement, because light creates its pathway. It shines the light in the direction it wants to go and moves accordingly.

There is no apparent aperture. The aperture is like a seal when it opens. It is a great archway that does not sit at the bottom of the craft, but rather to the side. It opens up and the light from within shines out. When it is open, a "tongue" comes out and the light makes a tunnel, or pathway, a safe passage into its context, or belly.

The ship of my people could be described as a domed triad.

The Beings.

The entities visiting your earth vary in appearance.

Some are very small and tiny. Remnants of them are buried near Suffolk, England. They once colonized this planet and give credence to the understanding of fairies and little people. Their ships are magnificent and yet you could hold them in your hand.

Then there are those entities that are a little larger, the more popularized variety. They are blue-gray of skin with large black eyes. They have no hair and no ears. They have only nose holes and a slit for a mouth, because they do not eat. They have no muscle fibre. They are made up of sinew, tissue, and bone. They eat prana, which is the ancient term for knowledge. They are coming from a system which has bred knowledge over and over, through their mother god, and they have bred out emotion. They are here to expropriate the human seed for breeding purposes, to bring back emotion to their genetic material. Tears are precious to them, and they are learning to love.

Then there is the variety that has come to do animal and vegetation investigations. They live on the interior of their planetary systems, not on the exterior. They have come to take home pieces of land masses with them, and to take the genetic seed to begin to cross with their own plant varieties to create paradise. They are the great gods. In the past, they have bred with human beings, and they seeded part of this race and many of you reading this material now. They are beautiful in nature; some exceeding over seven feet in height, with hair that falls to their backs. They are very broad and muscular. They have skin the color of cinnamon; it's often taken for golden skin. They have blue eyes and brown eyes. They have hair the color of wheat, and hair the color of night. They have been touted in legend as elves.

They are, in fact, the gods who blew away the ancient civilizations to create new ones. And their ships are brilliant white. Often seen up close, the smaller of their ships take on the appearance of a cut stone because they become transparent. When they begin to be airborne, the whole hull of the ship changes to a pinkish color and in a flash, is gone. They have the ability to disguise themselves in great land masses. They appear as clouds. You will often see them when there is nothing else in the sky except one particular cloud. They have inspired your motion pictures to communicate messages having to do with them and their beings. They are the gods of old. They come from the beyond the North Star. They are of my ancient people.

They are benevolent caretakers. They have been termed the Angels, the Archangels, the Overlords, the Constants. They are revered in truth. They are the ones that set religion going through the small ignorant minds of human beings. And to this day, they must appear as Mary or Joseph for you to be able to believe new truth. They wash away the blood and stench of ignorance to engage truth.

There are people walking amongst you from this group, and you are unaware. They appear common and kind. But they have enormous minds. They are spies! And they, above all others, understand the conditions of your condition. They keep everyone informed. There are some of you that have probes in your nose, behind your ears, and up your rectums. You are part of that system.

The probe lies up against the optic nerve and they see through your eyes. So how would they know when you ask to see them? The probe behind your ear let's them hear what you say. How would they know where you are? They know at all times. They know what you read, what you hear, what you are verbalizing. They know where you go. They are studying you. And seeds have the probes.

Let me make one thing clear. You mustn't create a new dogma around these probes. Please try not to make them the fashion statement of the decade. A lot of you already have them. But I can just hear you now.

"Well, how many do you have?"

"I have more probes than you. I am clearly the more enlightened!"

One day they will come out. You will blow your nose and there it will be. You will think it's just a grain of sand. But it's prickly. So then you will think it's a spore.

And by the way, it's alright if it comes out. It doesn't make you a reject!

What About Free Will?

Righteously you might ask, "What about free will? What about my rights?"

Well, we have all seen what your rights can do to this earth. What about your rights? What about them? How righteous are your rights to other people? Are you allowing? Or are you judgmental of other people? Are you angry or slow to anger? How righteous are you to the earth? What are you doing to make this a greater place? Is the earth blessed with your presence or damned by it? How righteous are you?

What about your free will!

Is it your will, truly, truly, to destroy the whole world? Is that the extent of your free will? Or is your free will only incumbent upon your own reality?

It's only incumbent upon your own reality. Because your free will isn't some blanket of truth that all others should abide by. And you, by no means, have the mind to determine the absolute knowledge, because you are bigoted by your own prejudice and your own fear.

Your free will extends only to the greatest of your judgment to determine what is real and unreal, after that it doesn't exist anywhere, any more than your past does.

What about your free will to work with these beings? One of the reasons you picked this up was to gain knowledge. They take a gamble on an individual's persona, their psyche, their ability in consciousness. When they contact you, they take a gamble that you have a potential for evolution.

Are they any different than those of you who tag animals and watch their migration? And are they any different from you when you move a nesting hen off her nest to look at her chicks and fondle them? Oh, you're careful to put everything back in order. But are they any different from you when you do that to nature? Where is the free will of the hen? And what chick gave you permission to fondle its body and its fur? Did you ask? NO! No, you didn't ask. Because in your mind, you're doing what you feel you have to, out of love and understanding for nature. That excuses it all. And they didn't ask you. Because in their mind, they are doing it for the preservation of the species.

Free will is relative.

But What About God?

What about God? Is God the gods? Yes. Is God you? Yes. Is He the hen on the chicks? Yes. Is it a He? Yes. Is it a She? Yes. Is it an Is? Yes. Is it the Life Force? Yes. Because what is it that allows the Is to be? That is what those great beings love and what they term knowledge. It is the pursuit, in consciousness, to make known the unknown. Don't you know that's their journey? And are their paths so different from yours? Do you not seek an adventure that will bring you joy? And are you not seeking that adventure out of the unknown? That is the same unknown that a brilliant ship can come from? It's the same unknown.

It is the same unknown.

You're asking for a lover out of the unknown, and they are asking for knowledge. And you're both getting it from the same place called God. That is one penetrating mind that is the all. When you are desiring small things in your world, they are consuming large ones in theirs. Where do you think the light comes from? Where do you think they got it from? From the unknown. From their consciousness. Are they any different from you? No. It's their degree of evolution that's different, by a large scale.

Are they any different from you when you try to save your flowering plant by watering it? They are trying to save you by getting rid of the poisons in your sky. Are you any different? And the god that "Behold God" is, is the IS unrealized. And both your journeys—a brilliant ship that would light up the entire night sky and a solitary person wanting awareness—are the same. You're relative in the same way.

What brings you together is the same realization. It's the same source you dip into for your own wants as they do for theirs. And it's that source in consciousness where I say to you, "ride the light," that brings you and all the travelers on the light together. You equalize, self to self. You are no less than the gods beyond the North Star who stand eight feet tall, with illuminated skin the color of golden cinnamon, with eyes as blue as the sky or brown as rich sable. You are no different from them, though their hair cascades beyond their shoulders in silken array, falling over formed muscles, massive and beautiful. Though they are ageless in their face with broad jaws and wide brows. You're no different. It's just a matter of different realities.

You look at you, unlit, and think you're not worth saving. They look at you and see God.

All of these are just words. But strung together, they create wonderful knowledge for you to consume. And these words were created to break down some of those barriers that your fickle little mind keeps creating around you. Knowledge is like light, it dispels the darkness. And there is knowledge available to dispel a lot of your darkness, to enable you to see the light. And you'll see it! You will learn to get a hold of your mind, start thinking, and *be* the thought. Be that hand that reaches out and gets the burger and then washes it in the clear water.

It's *being* this truth, this mind, and not being stagnated beyond an image that is afraid it's going to get bruised. There is not one of you that is greater than anyone else. And there's not one of you who's less great. There are only degrees of reality in this world. That's all. And those are marked by degrees in consciousness and that's it.

Some of you are reading this, but thinking of the guy next door. And some of you are thinking about what you should be doing now, or what's due at work tomorrow.

Others are simply wanting to understand this, plain and simple, so you don't have to run about like frightened, ignorant pigs running amok when you see something different in the night sky. Some of you desire to be knowledgeable people who are beginning the journey of consuming truth for a definite and real purpose.

This is not for show. This is for the purpose of locating seeds. And, by God, I hope you're worthy. Because if you are, you'll see a new world and a new time. And all of those old taboos on science will fall. When all of the old taboos against questioning minds are washed away and the tyrants are plowed under, the knowledge will flow like a river again. And you, in your lifetime—not in some other lifetime, but this lifetime—can be a precipitant and a participant in this experience.

There is a lot of contact to be had, and hopefully you will not encounter the hypnotizing ones who leave you with fearful dreams, but instead that you draw to you a remembered experience so that you can glean a whole new understanding that will preserve itself within you for the times that are to come.

How do you evolve without the friction
of life?
How do you evolve without breaking down
matter and changing energy?
You've got to live it!
The true master waking up is not the
philosopher. It is the alchemist
creating change.

Chapter 6

The Rim of Reality.

People say they want to grow spiritually, but they don't have any room in their reality. Imagine personal reality as a circle, and in this circle, there is an image. The image is broken up, sort of like a lot of puzzle pieces.

Your reality is like the circle, and the puzzle pieces constitute who you are.

When people begin a spiritual journey, they think they are going to go somewhere in a linear fashion. They make surface changes. They go to see their gurus and their teachers. They study philosophy and read all the latest books. Some people go all around the world looking for truth, looking for the important piece to the puzzle of their life.

This circle is lying so that the outside rim of that reality contains all of the things that are in your life, all the knowledge, all of the dogma, all of the things that are relative to civilization.

So the circle turns sideways, and you think you've begun a spiritual journey. You go out and you experience it all. You change the way your dress, and you change what you eat. (Well, altering the diet is an important part of the dogma, isn't it? Changing what you eat is only reversing your previous dogma!)

All you've really done is go from one side of the rim of your reality to the other side. That's all. You've just switched views on the same old dogmas.

You change what you wear. Instead of crucifixes, zircons, and amulets, you wear crystals.

Instead of repeating the Lord's prayer, you repeat your mantras. You've only gone from one side of the rim to the other side. Nothing has really happened. Everything you've done is surface.

And the image in the middle is still intact.

You don't ever seem to really change and go beyond the recognizable doors of civilization. You don't go beyond what "they" say is acceptable or normal. You're not supposed to eat certain things, "they" say. You're not supposed to wear certain things, "they" say.

You can be a religious fanatic, that's acceptable and normal. But be careful what you wear, and who you spend your time with. Someone is watching you! And be careful not to tarnish the image civilization has ordained for you.

But civilization is beige. It is made up of non-entities. It's made up of numbers on a computer. It's not made up of individuals.

So the concept of going on a spiritual journey just takes you to the other part of the rim of your same old circle. But the image stays intact because the truth is, you are very slow to undertake anything that would make a drastic change in your life.

All you're really doing in your spiritual journeying is switching roles. And there's nothing wrong with that. But I always find it very eyebrow-raising myself when someone asks incredulously, "Did you really change?"

As long as you stay within the rim of the circle, you can be a "spiritual" person. But you're not really supposed to make drastic moves in your life.

Few people even realize they really can change! Don't you know part of the reason people are "seeking" so desperately is to avoid being beige? Human beings want to understand their individuality, not the image of a civilization, and not the image of their *spiritual truth*. They want to *be* it. They don't want someone else's spiritual truth varnished on them.

You do not gain bliss until you start removing those plugs in your reality.

Think about it. If your reality, your eternity, your forever, is constituted by a circle, and the whole of your consciousness constitutes this circle, if you plugged it up with your image and what is acceptable to the civilized being, then you have no room for change. You have no room for the light to shine through. You have no room to be the miracle worker. You have no room to have the great thought that changes everything. You have no room. Because in order to make room, you have to change. And in order to change, you have to remove pieces of this puzzle to let the light start to shine through. That's what brings you bliss and joy.

I didn't come to create another part of the rim of your circle, but to blow out pieces of your puzzle. And that means change and change equates the unknown. When you blow out a piece of that image, there is not another piece of image that comes in its place, and the light starts to shine. There is a void there, but the void is joy. For a moment you have it!

I'm not liked a lot because I invoke change in people's lives. And I move them beyond their civilized acceptability. But you do not become God being a civilized acceptability and a number in a computer. And you do not know bliss when you have no room for it in your life. You are too sterile. Too rigid. The image is sterile and rigid. It says all the right things. It goes to all the right places. It does all the right things. And that never made a profit, not in what matters. No. And indeed, that never made a Christ. And it never made a legend.

No, a lot of people don't like me, because I invoke change. And I don't keep things intact. I blow them out. But I don't have to excuse myself to anyone. I only love you. And yes, I press you hard. And yes, I am an unrighteous dude. I never said I was Jesus Christ! I never pretended to be. I am underhanded and overhanded, and I will do anything to bring you to a point where you blow out pieces of your puzzle, so you can see the light of God, so that you can feel bliss! My God, when you touch superconsciousness for a moment, you know bliss, because you have joy in your soul. You can feel it! You don't know why, but it's there. Do you know what that is? That is another part of the puzzle blown out of your reality and the light is shining through. That's why the joy comes up.

When you are neither being a hypocrite to life, nor abating life, but living it on your terms, that's when you're real. When you are not a number, you're an individual. That's when you are the most earnest person on this earth. And that's when you know joy because that's what you desire.

And others may say of my people, "But they gave up so much."

But I ask you this. What is so much in the face and light of all eternity? You can never give up what doesn't belong to you. Your reality expands when your consciousness expands.

Yes, you have to make changes. You have to slide into the unknown in order to know bliss. And the only way to do that is to make the changes that blow out those plugs in your reality that allow change to come. You can't slide along on the rim of your old circle, because you'll just meet spiritual truth in a different form over on the other side. But meeting spiritual truth in a different form over on the other side of the same old rim doesn't *do* anything!

Because you haven't *really* changed; you aren't really living your truth. You keep walking out into the same old highways and byways, and you still act the same.

There is nothing written anywhere, not even in the ancient information from these teachers of 10,000 years ago, that teaches anything compared to what the God inside has in store for you. They could only indicate and be an example.

I know you've gone through a lot of fire. You have endured a lot. But any time you break down an image, and any time you break down the color beige in civilization, you are going to go through fire. Any Christ that ever lived and still lives, went through the same. But that was their chosen destiny. They wanted to evoke it. And yes, they had fear. And yes, they had wonder. And yes, there were times they wished they weren't who they were. And yes, they doubted. But there was something in them that drove them forward. They were legend. Because they lived as individuals. They didn't live as a part of that beiged-out rim of spirituality or civilization. They lived it. They were the fire of truth.

All of this "stuff" works; you just have to be it. You can't be separated from the truth. It must *be* you. If you have to strive to *know*, then you are separated. You will always stay equal distance from your striving and your knowing. Because, in consciousness, that is how you create it.

Most of the people who are in my consciousness are in a safe place. They have made changes because they wanted to. And yes, to the world those changes seem incredulous.

But the world is very incredulous right now. And that's what it takes to find bliss. And that's what it takes to find simplicity and the love of self, which is the love of God. And that's what it takes to have genius and a life filled with abundance and virtue. You are never going to *get it* living in the image. You will only grow old and die.

But, hey, you will have been, at least, civilized! And people could say you were a seeker. Then someone will bother to ask if you ever found it. Probably not.

Once you understand joy and bliss, you know the fire is worth it. I know a lot of people who are very happy. And yet there are those who would sit back and look at them and wonder how they could be happy. "Look at everything they've given up. How can they be happy. They live in the middle of no where!" Those questions exist because those who are doing the wondering haven't made changes themselves. They can't understand because they haven't gone fully through the experience to know. But once you understand joy and bliss, you know it's worth the price of blowing out a few pieces of a puzzle that didn't really ever serve you anyway.

You can say a lot about the Ram. And lots of people do! But there is one thing about my people. Those that are rich in spirit are not poor. And they are not weak. They are columns of light. And they live their joy and they are a blessing to those who happen to have them near. Because that bespeaks God and, indeed, Christ.

Bliss and change.
They are one in the same.
A master has the ability to glean the wisdom from every experience and the courage to move on.

Chapter 7

Bliss. Change. And destiny.

Breaking down the rigidity of who you are is what being a master is all about. The term master is a glorified spiritual term. Many people have all sorts of myths about what a master should be. But how would they know that? They are not the master.

A master is one who, with diligence and perseverance, engages his limited thinking in the light of knowledge. He engages the limitation to gain knowledge. He masters his dogma and his pet peeves, which make up that image in his reality. What he is doing is gaining the understanding to blow out pieces of his image, because every time a piece goes out, the light of understanding comes in and that enhances his reality. That empowers him. You are only as powerful as your consciousness. And if your consciousness is your image, then your power is very limited.

Now, the ancient teachers learned a mastership to explore a greater mind, and they mastered the things that inhibited their growth as a result.

It was purposeful to look at the understanding of bacteria and civilization. Both are the life force. And when you really understand that, then you can embrace that life is eternal and so there is no destruction.

The little beings with the large eyes were looking for a breeding colony, which they've gotten. Through taking back home the semen that you waste daily at times, and your unfertile eggs that wash out, and cross-breeding them with their own genetic forms, they are creating new people. They are endeavoring to breed in emotion, genetically.

They want the ability to weep. They want the ability to laugh. So, they are simply taking from you what you spill regularly. And you cry out, "Free will! How could they have molested me?" Well, you do it all the time. It is nothing new. Sometimes you molest yourself daily! All they are taking is what you didn't want!

When you stop contemplating their physiological form and their cohabitation on this planet and start looking a little bit further than fear, maybe the next contact won't have to be so frightening. Maybe they won't have to hypnotise you or paralyze you to protect themselves. Remember, they are very fragile and you are very fearful. You live in this "rim" of fear, and superstition, and anxiety, and hatred, and war.

You are warlike people. If you're not warring with your battle axes, you're doing it with your tongue. It's the same thing.

They hypnotize you because if you were to grab them, you would break their bones. They are physically very tender. They do what they do to protect themselves.

There is something else you might like to know. Those of you that have these little prickly pear implants up your nose and elsewhere have given the beings that implanted them a great deal of understanding about your way of life. They see through your eyes. They hear through your ears. They understand what you consume. They know how your gaseous tract works. They understand you. That's what they wanted to do. They want to know how you have evolved since they were here 10,000 years ago, when they pulled up stakes and left. These people aren't here to entertain you, but to do something about this place, because this place is in a very bad shape. I know you don't know that, but take my word for it. It is a truth.

They are here again to bring the teachers, to engage a civilization, to pick out the seeds, and to plant the truth. They are here to herd the seeds to a place. They realize everything is about to break loose here and they're going to help in it. Because they understand what is needed and they understand the imminence of destruction.

When you have a 500 megaton cobalt bomb in the hands of your enemy, or your best friend who could turn on you any moment, then the world is in a perilous shape. Not to mention, of course, the residue of all the people who left the land and went to the city. Someone has to deal with all of the greed and pollution this world is suffering from.

So beings are here to engage this planet and to choose their people as seeds, as they have always done. And your god, your spirit that is equal to them, has given an agreement to do so.

There is an armada here on your behalf.

When they destroyed these civilizations (and there were more than I have told you about) in the twinkling of a eye, they made decisions based on situations very similar to what is going on in your world today. The only difference is that in the past there was warning after warning, and no one heeded the warnings. Here in your society today, the truth of the state of your earth is hushed up.

It's a big secret that the world is dying. Did you know that? They can't let the investors get hold of that kind of information. They've got to keep the money coming in and interests rates have got to stay high and the dollar has to stay the same. The dollars are what's important. You can't let the truth out because your investors would be enraged.

The dependence on your paper money illusion is so great that they would hide the fact that the world is dying just so that check can come in every month. That's what you have when people leave the land, leaving behind only a few responsible for providing food, clothing, shelter, water and the bounty of life. That's what happens when humanity chooses to live in caves made of glass instead of in harmony with the earth. When you are no longer in harmony with the earth, you are no longer in harmony with the life force. You get out of sync.

Nature is ahead of the human being in evolution, and yet you call this civilization. Civilization is one of the most dangerous things that ever happened to the human being.

The process of civilization is a learning, yes. But it is a trap. Look what you have to go through just to be acceptable every day. Look what you depend on. Can you stop driving your auto machine? Can you stop turning the key and spewing exhaust out of the other end? Can you stop buying petrol? Can you stop using plastics?

What would people think of you if you hitched up a horse and drove downtown? Novel idea! But no one would like you if you did that, and you couldn't face the ridicule. And yet, what that horse eats comes right off of the earth, and what comes out the other end fertilizes the earth. It's in harmony. That's a very crude example. There are people living on your plane today who live like that.

The aliens that are here have, by and large, come to help get things in order because what affects this earth affects everything in this galaxy. This galaxy is sort of like your body. Let's say you get a sore someplace on your body, someplace you hardly ever notice. All of a sudden, all your focus goes to that sore, doesn't it? If you have an itch, all of your focus goes to that part of the body, doesn't it? All of your focus goes there to relieve it.

The whole galaxy is like your body and if there is a sore breaking lose, all of the attention, all of the awareness of that body, goes to that sore. And guess what? The sore is you!

I am telling you this because the more you begin to have knowledge pouring into your vessel, the greater your consciousness is going to be, and the greater your consciousness, the greater your awareness. This is what magnetizes to you the same idealism and it is the one thing that will preserve you as well. It's that light coming through the pieces of the puzzle of your reality.

Every time you see a green fireball burst in the skies, it's endeavoring to neutralize the poison sitting there. And this isn't just for your sake. Those creatures that live in the sea have a soul. The animals that are dying have a soul. And they have a spirit, and their outcry is also being heard.

The great teachers of 10,000 years ago, that brought knowledge to all the civilizations, are back again. The teachers that moved the Eskimos from Ceylon to the frozen North, and transported beings around and held schools, endeavoring to teach human beings to get them beyond that destructive gene, are back again. The teachers that have endeavored to turn around the human drama are returning to expose what the world is trying to keep a secret. That you are dying just so your checks will come in every month!

We've moved away from that magical concept of aliens who love you and want to beam you up and take you home because you are so sweet. We have moved away from that romantic notion that somehow you are going to teach them something! Hardly! And we have moved away from that idea that somehow you are great enough that they would want you. Why would they want you? They are your brothers who love you. They are minds. They are knowledge. They don't have this image problem you have. They are evolving, just as you are evolving, but they aren't controlled by an "image."

They are taking action as ones who love you, in order to help you. You will make contact with them when you are of same mind and same purpose. When you get over the concept that you are going to be swooped up to some great ship and taken away, when you get over the concept that your soulmate is on one of those ships, when you get over the concept of all of those ideas that are on the aforementioned rim of life, the image problems, then you are going to have a great contact, because then you will have magnetized to you the same.

You magnetize through your auric field what you create in consciousness. What you expand in consciousness, you magnetize to you. That is reality.

That's why things fall away when you change. Because when you change, you have gaps in your life and joy can come in. When you change, you get happy. You don't know why, you just do. What you have in consciousness, you magnetize to you. That's how it works, people. It's so simple.

It's all relative.

What you are learning is not going out the back door. It's not. Just as certainly as you have made changes—and those changes have happened as the result of consciousness—it is possible to magnetize a partnership with these people. It works. But you have to prepare yourself to be a partner. You can't be a funky, spiritualist partner. You have to be a real person. We are not looking for image, we are looking for substance. We are looking for God and the ability to unfold that, to be that, to live that, to express that.

They don't pick nincompoops. They don't pick you based on how famous you are, or how rich you are, or how poor you are. They pick you on your simplicity, and they reveal only to that. Because in that simplicity is brilliance. When you start living your truth, in your life, you magnetize same cause to you.

Don't do this to make a name for yourself. Do this because you love what you are and because you love the life force. (The life force is the light that you ride when you get out of this body.)

These people have a will of their own. Haven't you ever had friends who didn't show up for dinner? These entities are watching. They know exactly who you are. They know exactly what you are doing. They are very aware. They don't have vision like you do, but they are very aware of you.

Some of the sightings you have are really greetings, sort of a hello. When my daughter was first introduced to the Ram, I had to send a runner to convince her there was such a thing, and that I was not, in fact, the devil. The runner was a wise old woman who my daughter went to see. And, in order to convince her further, because she needed a lot of convincing, I sent a message through a Ouiji Board. (Yes, I am unscrupulous!) And the message said, "Oh, wise one, we salute you. Greetings! Come outside that you may see the lights of heaven." So, immediately they ran outside and there we came, as great lights in the sky. We came overhead, split off, went up to the highest zenith, turned, and put on a spectacular show. And it was just to say, "Hello. Yes, we exist." Just to do some convincing.

And someone said, "It could have been a meteor!"

It was planned much too wonderfully to be a meteor.

The lights that you see when you stare at the night sky are a greeting. They are the first "hello." You will see a lot more of them. And if they appear to be shy, they are. And if they do appear to you and you are appalled at the way they look, you may lose the greatest bridge of light you will ever have in this life. But if they come to you, in radiant light, and your God rises up within you in absolute love, you will have an adventure beyond any words I could find in your vocabulary.

You are taking the time to learn, to turn your subjective mind to an objective mind. That's very difficult but that's what it takes to look at a being who looks nothing like you. You are so conditioned from horror movies and superstitions that you consider anything different from you to be evil and bad.

Objective minds sees it differently. In consciousness and energy, you reach the objective consciousness. It's called the unknown, and that's when God will rise up and engage.

This information is as factual as breathing. I want you to know these beings are as real as your friends you see every day. In fact, these entities are quite like your neighbor next door. If you don't make an effort to show warmth and love, then you are not going to be able to engage relationship. You oftentimes have to take the first step. And yes, I know you are earthbound and you can't zoom up into the sky, but one day you will be able to levitate off the ground and it will be quite a feat. Those sounds you make when you are pumping air in the process I teach called consciousness and energy is how the masters in the monasteries levitate.

You are making gestures and they are saying "hello." When you bless them, that is a greeting.

What you need to learn is how to grow beyond the puzzle of your life. Because the purer you are in consciousness, the greater the contacts. And if you're doing this to show the world you've had contact, chances are you aren't going to have one. But if you're doing it from the love of the God within you, because they too live on this consciousness, then the relationship will be a very wonderful one. Now do you understand why it is important not to take this subjectively? When you no longer take this information subjectively, that is when you get into *great mind*, and that is what they are.

Have you noticed a lot of people are moving? People are being moved by spirit. They are becoming aware and they are responding, even though there aren't any angels appearing at this time telling people to be aware.

I want you to know it speaks of who you are, if you have heard the call and you've listened to your spirit. If your spirit is moving you in harmony with this truth, with these words, then you are part of the seeds that will see it all and outlive it all. And that's when we can once again open up communication and that will be wonderful. And the teachers of 10,000 years ago can return. And the creators of the tombstones to the living can return, the creators of those 20 meter high, 50 ton statues staring out to the sky, mutely saying, "We are coming back. We're not buried here. We're out there, where the eye gazes. And we're coming back."

Communication is opened when consciousness is ready for it. There have been civilizations that have been ready for it in your history. There are civilizations that weren't destroyed. They simply left. They evolved *out* of this understanding and they are gone. They are in a different dimension. That's their natural state of evolution. And there are civilizations that have been destroyed in the twinkling of an eye, as I've mentioned.

Consciousness has to be ready for an engagement with those who come from beyond the sun, and you're on the edge of such an epic adventure. The awareness is coming from everywhere. Those who can't see it are dead in spirit. They can't change. They have to live by the image. They are what you call "civilization." The few who see it and understand it, will begin to bridge these wonderful teachers and the knowledge they have to share with you.

These teachers would not come to exalt themselves, but to exalt you, as equals.

Your image can do one of two things with this information. It can "blowout," because it gets overcharged; or it can brush it aside and say, "That was an interesting story, but it will never happen." In which case, it never will. Because you determine how much of this truth you want in your reality. Did you know that? When I tell you this information, it is simply words. When it becomes your truth, then it magnetizes that reality. Then you are a living truth!

There is a legion that is waking up and it has taken that in every civilization.

There are people who are finally waking up to the destruction of the planet and there will be an outcry. That is a blessing.

And someday humanity will wake up off its soft haunches and begin to march against tyranny, and when that happens, tyranny won't be able to do anything but be overthrown. Look at the numbers. Only a small group of people control the world. What if everyone woke up and said, "Enough. I'm not going to have this in my experience any longer." Then you would have the first civilization that wasn't history!

History repeats itself because civilization is an image of the past. America is built on Rome's ashes. That's how it works. Every civilization gets "beiged" out, hypnotized out, civilized out, until they become ignorant numbers. You're not freethinking people, and you're kept from being freethinking people through religion! That's the sword of these times.

If everyone woke up and said "Enough!" and took up the march against the governments, against the lawmakers, and against those who control the greed, things would change. If enough people rejected the plastic and said "Give it to me in my hand." If enough people turned off their electricity, refused to use the service, and refused to turn on the car, then genius would run rampant. Because sitting in the files at the auto makers are the plans for perpetual motion machines; they just couldn't "afford" to make them before.

But if energy were free, how would they control the people then? And we certainly can't have that!

Don't you know what you are in reality? I'll tell you what your teachers have known. You are slaves under the yoke of a tyrannical few who rule the world. And you say "OK" every time you write out a check!

There are many things you can affect, but there are some things that cannot be changed. There are some things that only geological time or supernatural help can take care of. And one of the things you can't change is the state of your oceans.

You cannot clean up your oceans. What would you clean them up with? Do you have a vacuum cleaner that's big enough? Do you have a sieve that large? Do you have enough chemicals to counteract what you've thrown in there? Even if you had enough chemicals to counteract what you've done, the chemicals would make the oceans sterile, and it would take another 100 million years to get life forms going as you know them today. What are you going to do? Do you still like to eat your fish? You are eating your dung. No, in reality you can't dump counter-measured chemicals into the sea, but you can do a lot of other things.

These entities that you want to contact are here on big business. They are the only ones who can do something about your oceans, and what they have planned is not for you to know. You only need to know that is where your help is coming from. It can't come from human beings. It's too late for that.

And did you know your sky is falling? It's tearing, and it's falling. I could tell you some other things, but I'd be accused of spreading fear in the marketplace.

Sometimes it takes shocking realizations to wake people up!

We must focus on the next decade. That's what you are changing now. When I have said "You are in my consciousness," know that I meant it. The few of you that I said are the seeds of knowledge are just that. And I'll pour the information to you so that you are not ignorant, so that you are not left walking around in that civilized image without the truth. I want you to know how to command that truth and how to live that truth, so that when you have contact, you won't turn around and run away. I want you to have enough knowledge to have the courage to engage. And you won't have to run and tell the newspapers just to exalt yourself. (There are people who do that, just as there are people who push causes just to exalt self.)

Be what you are and move forward.

Take this knowledge, expand your consciousness, and align yourself everyday. If you can blow your nose, you can do Consciousness and Energy. You don't have to make time to do it. It should be the thing that destroys your concept of time and creates the *now*. You want to see more things? Expand your consciousness so that you *can* see more. Open up that little part of your brain right behind your ears. There is a section of brain tissue there the size of a nut. When the pineal is thrust open with the movement of energy, there is an awakening of electricity, as you would call it, that surges to that part of the brain. This is the most sensitive part because it allows the "visual" to see interdimensional. (It will allow you to see me.) The process of opening up that part of the brain allows you to see light. It allows you to see what is "not seen." Every day that you do Consciousness and Energy, you are pumping and expanding consciousness to engage the unseen.

And don't tell me it doesn't do anything. It does. It works. It blows out and creates joy. And if all that it gives you is a moment of laughter, it has done a wondrous thing because that laughter will magnetize that same sort of reality to you.

In the next decade, you have a forward march through turbulent times. You have to dig in and continue to make yourselves sovereign, to be free. And free doesn't mean a "mooch." Mooches are leeches. Leeches aren't sovereign.

You must have your own place. You need to "dig in" and get stable because once you've got all of that taken care of, you can pull in the knowledge, and it will come to you like a river.

Bliss and change. They are one in the same. And a master has the ability to glean the wisdom from every experience and the courage to move on. That is bliss.

If you can't glean the wisdom and the courage from every experience, then you're caught in the groove. And that sliver of light shining through the puzzle gets pasted back over with civilized imagery, and you go back into the groove. And you're never the same person. Ever.

Bliss is having the grace to discover a moment that is without the illusion of the past or the future. It is a moment collapsed in time to be that moment. Not everyone is ready to *own* the image. Not everyone is ready in their evolution because not everyone has the seed of love and knowledge, which gives you the courage, which allows you to address these things. That doesn't make anyone wrong or backward. It's just the way it is.

These are hard teachings because they evoke change and that is equated to pain. And most people never go all the way to the other side; they stop right in the middle. Because it hurts too much, and they run back, and never gain the full extent of the experience. They never gain the wisdom. They gain only regret, pain, agony, and sorrow.

My people are beautiful entities. More beautiful than your eye can perceive as beautiful. They are the fiery chariots of old.

And yes, they still give the vision of the Virgin Mary to get a message to the church, trying to change the sucker. Do you know how many Mother Mary's there have been? You have to do that with people. You have to give them an image to be believed. Some of you will see something that doesn't make any sense to you, and that's because you aren't capable of really seeing what is giving the image, because it would frighten you. You would shut off. So you have to see things in *imagery*. The Book of Books is filled with imagery perpetrated upon humanity just to get a message across.

Those great gods are still around. And yes, they too might be considered perpetrators. But you must remember they are dealing with a mentality that is very small and superstitious and ignorant and afraid. So they have to address you with truth, according to your ability to receive it. It will be a wonderful day when you can see them without the image.

Remember this. They, too, have a destiny. And theirs is a destiny of evolution as well. Their evolution cannot even be comprehended with your linear mind, because you only understand things from point A to point B, linearly. Their destiny is to *be* God, to be all they can be.

It is not within the human nature to be bored; boredom is death to the human spirit. (We are speaking of humans who are interdimensional and interstellar and those in parallel universes as well.) That is why civilizations are historically "history," because they repeat the same things, because they stay caught in the image. They never evolve. These people are evolving. They have their own destiny.

Now, let's take a look at you. You are earthbound. You don't have a fiery ship for riding the night skies. You don't have wings, and you're not seven feet tall. You don't have illuminated complexions. But you're still God. You are as important as they are. Never see yourself as less than anyone. You are only in a different stage of evolution. Your destiny as God is just as important as their destiny. You, the human being, have a role to play; and your spirit presses you to play it.

I didn't have to come here and deliver this information. You have no idea what I go through to come here. I know some of you wonder if I am a worthy teacher. Have you ever wondered what I think about you? Are you worthy of these teachings? Are you worthy of love? But there is something very important about you that I love. It is simply that you are God, and that, I love.

You have your own destiny and you will make it whatever you want it to be, according to your spirit that demands that you evolve. And that is a righteous thing, because your next life may be where these beings are now.

Maybe your next life will be far away on a planet with a blue sun instead of a yellow one. And who's to say that this current life of yours is the dredge of all of your lives. It is a step in evolution. Your life, seen linearly, would be seen as steps through seven levels, and each step would have a footprint on it. And that footprint is each life, or lives collectively, that bring forth illumination. And as you take a step, the next one will be greater yet. So this life is one great step in your evolution.

What is the next one? They only exist in the moment you create now. That is what is important—*now*. Don't ever look up at those people and wish to be them. If you do, you devalue self. Don't ever wish that you were living somewhere else out there because if you do, you destroy your reality. Be who you are and evolve. It will bring you greater results and more spectacular adventures than your mind could have ever perpetrated upon you.

The bottom line is, you are worth it all.

You're all part of a great tapestry, a great plan. You're part of the threads that go through that tapestry. And you may not know it, because you can't know all now. You will evolve into knowing it all, but what you're learning here is very important for times that are coming.

There is nothing in the image
that will ever make you happy, people.
There is nothing in the past
worth remembering.
There is nothing in the future
worth hoping for.
There is only this moment.
And God.
And realization.
Because that is eternal.

Chapter 8

Consciousness and Energy Create the Nature of Reality.

I have not said the world is going to end. I have said the world is dying. But it has help. It has big time help.

Your world has to change.

We spoke of the image and the puzzle in your circle of reality, and how it has to be blown out. Well, the world is quite the same. It's the life force, and it has to change as well.

These great brothers, these ancient teachers, are here to participate. And you are going to participate. You must, or else you will fade into yesterday.

But I didn't say you wouldn't participate or fade into yesterday. And I didn't say the world was going burn to a cinder. It could, very easily, your atmosphere is very volatile. The people who hold death—the people who hold the distribution of matter—are in a very volatile situation. Things are very volatile.

But in changing consciousness, in superconsciousness, a new heaven and a new earth can be brought forth.

The reference to heaven here means the firmament, the nebulae, and the new earth means a changed person, the mother being changed.

Let me clear this up because I know how you take things literally and out of context. The world is not going to come to an end. Civilizations will. They are already rotting away. But the world will have to change.

What you are learning about interdimensional mind is very important, because in order to be a part of superconsciousness and the new world, you have to have an interdimensional mind. You are learning to have an interdimensional mind in consciousness. And as your consciousness grows and you gain the understanding of being objective, even more knowledge will be poured to you.

Many of you are in my consciousness, not all, but many. And you are going to be fed illumination. You are making yourself worthy for it. And that, in turn, magnetizes to you the experience of that illumination.

Many of you are seeds that have to grow. You are part of a great plan. And I know it is very easy to waltz into the image again when you feel like you're so small you can't make a difference. I've seen you talk yourself into feeling small and overwhelmed. And I've seen you begin the process of change and expanding your consciousness, only to get frightened, because as things begin to change in your life, the things that identified you start to fade away, and you find yourself feeling helpless. You reach a place where you don't know how to describe you anymore. You may not know it, but that is a wonderful sign of your illumination!

I understand that when you start to work with the process of Consciousness and Energy, you think that the unknown is also unsupported. You must understand this. That is a myth. The unknown is there to be consumed in consciousness. It supports you because it is what makes life the IS. You must consume the unknown in order to be expanded from it. You are being supported by your God.

And yes, I know you look for signs. But don't you know that signs only resemble the past? Because in order to support you, you would have to have an image from the past. The void, in and of itself, is the sign. That is the unawakened mind awakening. That's what it takes, and that is an alone journey.

You look at this process I use to communicate. You hear a voice, and you see a body, and it helps you to identify what is being said here. But there is something behind these eyes that loves you. And it is powerful. And it is with you all the way into that void. Because I have been there and back. I know where you're going. I know what it takes.

And one day when this tapestry is completed, you will be able to see the whole picture and the remarkable part you played in it. Because you will be alive to look at it!

When I said that to have interdimensional understanding required an objective mind, you need to understand we are working to have an objective mind in a very small amount of time. There are entities who study all of their life just to try to get rid of their image. There are entities who spend all of their life learning to breathe, learning to be objective. We are collapsing that, and endeavoring to accomplish objectivity in as small an amount of time as possible. We've had to make an appointment in your reality. And somewhere in that small little rim of your reality, we are working to turn around a whole conditioning, a whole victimization, a whole subjectiveness of mind, to create an objective point of view. Because only in that way are we going to be able to expand consciousness. When you practice Consciousness and Energy, you are moving objectively. You are moving in consciousness. You are "blowing out" the image; you are burning it every moment. It's the only thing in the human drama that works to get you there.

I know there is a certain temptation to swing back to subjectiveness. That is point of view. There is a true temptation to move from the desert, so to speak, back to where you came from, because that is familiar in thinking and this seems alien. But we are discussing meeting an alien mind, so you have to create an alien mind. Not everyone will be able to do that, and yet it is the simplest thing to do. You just must be able to break through the rigidity of not just your body, but your thought processes as well.

You are making a difference. And you've taken the first steps to communication by taking the first step within yourself.

There is nothing in the image that will ever make you happy, people. Never. There is nothing in the past worth remembering. There is nothing in the future worth hoping for. There is only this moment. And God. And realization. Because that is eternal.

You can have all the fine teachings you want. You can even go back to old tapes and listen to me talk when I couldn't communicate so clearly, because I didn't know your new words. It sounded very mystical but when you strip it all down to the bottom line, I'm telling you the same thing now.

You can do it. You just have to want to do it. And all the hoopla and airy-fairy mumbo-jumbo just piles on layers to your image that one day you will have to address and break through again.

Simply expanding consciousness not only connects you to the kingdom of heaven within you, but also connects you to your god within you, to yourself. And that armors your destiny and takes hold of you, allowing spirit to move on the flesh, turning the pages in your book of life. And your god within will not let go until your destiny is completed, and you'll never have to want for another thing because that god will make it happen. It knows exactly what you need. You don't even need to ask. When you ask, you separate from it. You just have to *be* it. That's all.

The bottom line is, you can do it. In a moment you can have a mind that hooks up to universal mind. All the other stuff is just layers of garbage. Spiritual pollution. Recycled ignorance.

You will never be the same after reading this book. You will have the information to empower self and your divinity. You will see through the illusions. You will have the information to find the joy, and the strength to manifest the reality that is so needed. And you will have the grace to see yourself through it all. It just takes one little thread. God.

A few people have left my teachings. It wasn't what they wanted. That's the way it should be. Never do anything you don't want to do. Don't ever betray yourself, because that's what makes an image, and that's the most difficult thing to get rid of. That's why a master, masters all of his life for the moment of Christ. But when you want to do something, you do it. And you should move mountains to do it. Because that *want* is a necessity for self to evolve.

The world is matter. It is matter waiting to be transformed, transfigured into a greater energy. The human body is also matter, waiting to be transformed into wise men and wise women. That process of transfiguration is the alchemy of matter. When that transfiguration happens, it releases a divine energy. And just as matter is transfigured when the secrets of its energy are released, so is the body. The human body is matter waiting to be transformed and transfigured to release its energy. It is that energy of the wise man and wise woman that attracts the super beings, the Lords of the Cosmos. It is that energy that attracts super life. We are speaking here of a transfiguration. When you disseminate matter, you have a new life form.

An alchemist will live in abject poverty, using the same method, over and over, every single day to break down matter. It would take seven years, working every single day, to create that change. He will go without food, clothing, or heat, just to fire the same matter every day, waiting for a change. And he waits for it. Why would such a person wait to see the transfiguration of matter in his cup? Because if the alchemist can look and see the stars rolling in black liquid, then he has transformed matter; he has transmuted it, and he knows that means the very process was a transmutation of himself. He's willing to wait for that moment, because in that moment in his ritual of transmuting matter, he will be transfigured, transmuted, released. He will have been released from matter himself, in consciousness, therefore he can look at the whole of the world, and see its illusions, and be free, forever, and ever, and ever. That is his path. That is the same path a master takes. Transfiguration.

The world is matter waiting to be transfigured into a greater energy source. The human being is matter waiting to be transfigured into wise men and women that attract superconsciousness.

The transfiguration of self is the work of an alchemist and the work of the alchemist is the spirit. And the spirit will work, and work, and work in consciousness, to break down matter, so that it can be transformed, so it can be illuminated to release its energy. Remember I said, "Go beyond your rigidity. You are greater than your bodies. You are forever, and ever, and ever." That is an outcry to your spirit to release your energy. It is a command to your spirit, the alchemist, to transform, so that you are released. Why go to all this bother? Because it is the next step in a collapsed time spiral's evolution. You are moving rapidly over page four, five, and six in the book of life, and you are right on seven. That's what you are doing.

When you transfigure, you attract the Overlords of the Cosmos. When you transfigure, you release the energy that is like honey, an elixir to superbeings. When you transform, you release energy that is, in essence, a magnetism. On one end, the magnet releases this reality, and everything that is no longer progressive. At the other end, you are drawing to you, magnetizing to you, the great unknown.

The only way you break down matter is to transform it, and an entity who does that is not a sorcerer, but an alchemist. The alchemist within is your spirit, and it breaks down the physical to transform its energy.

In school, in the process of Consciousness and Energy, I push people to keep working, even when the body doesn't want to move and it doesn't have any breath left in it. I do that because it is at that moment of breaking down the physical that the spirit engages. That's the moment when you are transfiguring. That's when you are releasing energy from an unknown consciousness that is exploding. It is that energy that magnetizes to you all that is in the unknown, and it is that energy that attracts the super beings. It is that energy that attracts genius. It is that energy that attracts everything that you could never have conceived of in one lifetime.

So why does the alchemist arise every day from his tattered pallet, and have his meager pot of tea and his bread, and begin to work with his matter? Because he knows—and you should know—that in as much as matter can be transformed and its energy released, so are you.

Remember, you are one with the life force. What is relative, is relative to all. And the alchemist is waiting for that moment of his release. That is the moment of his Christ.

What are you doing to magnetize? You must make a conscious attempt at evoking your alchemist, which is your spiritual god, to transfigure self. You must break down the rigidity of mass, and rigidity of mass is held together only by your consciousness.

You are the one who determines if you are blind or deaf. You are the one who determines what your body is; it just follows the orders. But you are breaking down the orders. You are breaking down the physical in Consciousness and Energy. You are wearing it out, until you reach the point where it cannot possibly get up again to evoke the spirit, and that is the moment that the alchemist, the spirit, comes in. It is at that moment that there is a rejuvenation, a lifting. That is the moment that you turn heat to cold. That is the moment the body is healed, because a new consciousness comes into it; it is transfigured.

When you do Consciousness and Energy, you are evoking the alchemist, the god. When you do Consciousness and

Energy, you are evoking power, consciousness, to transmute matter, to break down the physical so that it can be changed. And in the process, you are breaking down the image, so that the burst of light of consciousness can come forward, allowing its energy to come through a new vessel that magnetizes to it, God.

You can sit and do deep-breathing all day long. As a matter of fact, you can do that for the rest of your life, and you will simply create deep-breathing lungs. You can sit and emulate what you perceive this process to be and it won't do anything without the knowledge. Because without the knowledge, the image is only glorified going through the process, because it becomes just another dogma. You can wear your crystals, your crosses, and your zircons. You can read your holy books and your unholy books. You can do your mantras—which became a hypnotist and froze the soul—and it won't matter. How do you evolve without the friction of life? How do you evolve without breaking down matter and changing energy? You have to live it.

Every moment you engage Consciousness and Energy, you are turning and grinding matter; you are grinding the limitation of consciousness. And it is firing. That heat coming off of that body is transmuting those cells. It is a breath of life. It is an unknown power surging out of you. It is healing. Every time you breath in Consciousness and Energy, it pumps the cells full of life. Your skin will get younger; your countenance will get healthier. There is no room for disease when the breath of life is being pumped to break down the flesh to change it. It's an aftereffect.

And the joy comes when the alchemist knows, after 63 years of work, he is coming to that final morning. He will arise one more morning, and he fires the kiln, and brings out his cup. And the matter will run like a blue rhapsody, and the diamonds of forever, the stars, shine within. And in that moment, he is gone. For he belongs to eternity. He has become the matter he has changed, forever, and ever, and ever.

The world is waiting, as matter, to be changed. It's dying. It has to be changed. The human being is waiting for the human psyche to be engaged, and in that engagement, be transformed. The human body is waiting to become light. It is waiting for a rebirth.

It is that magnetism that gives you eternity. It is that breath that is the breath of life that gives you forever. You must be transfigured and you have to change to do that. And you have to want to change. You have to want to. Because when you work with consciousness and energy, you create an expanded consciousness, and you can create a turbulent hell for yourself if you create something that manifests, but when you get it, you don't really want it. Then you have to deal with what you created. Because what you create has no judgement. It does not hold back. If you call it forward, it is law, and it must come forward. Everything properly created in your consciousness is magnetized. It will be your reality. It happens immediately when you engage it.

Manifestation is the aftereffect of materialization in consciousness.

Are you ready to engage those doors? Are you ready to watch things flow out of your reality because they no longer fit? When things flow away because they no longer work in your reality, are you going to turn around and grab them and bring them back? Will you be surprised that you can't communicate with people? Will you be surprised to have joy in your soul?

How do you think you get interdimensional mind? You have to transfigure to get that. You have to be transmuted in order to be interdimensional mind. And when it begins to happen, you'll feel it. And that's when you begin to rise off the floor. And that's when you're greater than your body.

That is the true master waking up. It is not the master of philosophy. It is an alchemist creating change.

This releasing of energy is what draws the superbeings to you. You see, it doesn't matter about time and dimensions and barriers.

What you magnetize is relative to what is. And when you release energy, it magnetically draws them to you, because you are destined to be together. That magnetism is relative, and it doesn't care about linear time, distance, space, or dimensions. I created Consciousness and Energy in my lifetime. No one has ever known about it. It would later become the lotus position with relaxed hands to receive. That was the result of a lazy person who decided the original was too hard to do. A lazy person decided to sit with relaxed hands to wait for the energy, or the kundalini, to rise from the spine. What do you think the kundalini is? It is energy coiled up in the loins. (It's the same truth; it's just that now it's a dogma, a myth.) And the only way it rises is when you evoke the spirit to call it forward. Self, the alchemist, does that. And no meditation in the mind will make it happen. There is nothing "out there" that is going to make it happen. There is no light, no crystal, no proper food, no proper clothing, no weather. It is the process of Consciousness and Energy that makes it happen. God calls it forth.

That is how I ascended in dimensions. I know about transfiguration. It is an aftereffect of self, realized. I am teaching you that. No one else on this plane knows this technique. Others imitate, but they do not own it. Because they don't understand the word transmutation. They don't understand the word transfiguration, because they are still separate. They don't understand the nebulous God. They understand only dogma.

The process called Consciousness and Energy is a science. It is one of the secrets. When you do it, you gain knowledge and your consciousness explodes. You release energy and become more divine and more magical every time you evoke the process. The process should be more important than your food. If this is the way to be god, and you want to do all the things the masters of old could do, and this is the way to do it, you should want this before your food.

It heals the body by breaking it down and restoring it and changing it. It will break it down to change its structure. It will heal it. It will keep the body young. It will never grow old and, more importantly, it will never die, unless you give it permission.

It pumps you full of energy that makes you lighter than air. You levitate, if that's what you want to do. This is how the monks do it. You can heat up things or make them cold. Temperature changes in your body will soon become an outside phenomenon; temperature will begin to change around you as well. That's the way it works. And you can wave your hand and things happen. But the prize is being transfigured, Christ/God/Man/Woman realized, all you can be. That is a forever being.

Consciousness and Energy burns the image and burning the image destroys the reality of the image. People won't understand you. Many people won't be able to communicate with you. Some people will leave you. And some people around you will be unhappy. Because you are going to make changes. And don't be surprised when changes start happening. When they do happen, don't start acting like a victim. You evoked those changes; be ready to engage them.

They are the next step. They are the morning of grinding and firing the matter. Do you understand? And don't be surprised when electrical things fail you. It just is. And don't be surprised when you become transparent; it just happens. It is part of the process.

It's not the phenomenon; it is the truth. You, being what you are, are the prize; all other things are aftereffects of that. But you must be transformed or else you stay in the groove. And another layer of syrup will go on top of that image of yours, all smeared with enlightened varnish and sticky meditations about love and the garden path. And, to make it look good, you'll put some crystals on it, and some pictures of your guru. You can put some pictures of Jesus on the cross there. (You should, at least, take him off the cross, you know!)

If you keep piling more stuff on top of your image, you'll never know God, and you'll never have an extraordinary experience. You'll continue to grow old, no matter if you are a vegetarian and all you eat is fruit and bird seed. You'll still grow old. (It's not what *you eat*. It's what *you are*.) And you'll still be unhappy, no matter how many mantras you say. And you'll still be a leech, because you won't be able to make your own way in the world any longer. And you won't be worthy of help because you won't be making an effort. You'll just be spiritual and divine and all of that. And you'll die. And you never will have magnetized greatness to you, and you will never be known as legend.

Not everyone is ready to learn the truth and those who learn this can only come that way by pure spirit. You cannot gain the power through the image. You have to break the image down, therefore this information is protected. Even if you thought you were learning this to destroy, you could never destroy, because you will never have the power. You have to be reborn, and being reborn is doing away with that little genetic chromosome, that soulful need for destruction.

So if you learn this process and sometime you're doing it and you think you aren't getting anywhere, you aren't.

You're just doing it for your image, not to really make a change, so you won't change.

Magnetism. To magnetize. To equalize. To be relative. Everything is relative. If you want great things, you have to become great things.

Can you ever go back to yesterday? No. Because there isn't a yesterday. And you know that and I know that. All you can do is create the images of yesterday for today. America is built on the ashes of Rome. History is today. You can stop this process of growth and still be an aware person, to a degree. But there'll always be this longing, and a hunger, and years will flow like a river when you hear the words "Behold God." That phrase will do something to you forever and ever and ever. And when someone says "You are great," it'll do something to you. And you'll have to live with knowing you chose not to go on to be all you could be. You'll know there is something missing in this life.

How great can your alchemist be? To define the word great you would have to say it is supreme. If it is supreme, it has no equal, so it just is. You just are, you know. How far you want to take this is all up to you, of course. It is always on your terms.

Does this mean you have to give up things you don't want to give up? No. Don't ever do that. There should be a natural evolution to dispatching things in your reality. But don't give up something that you love and want to keep, because that means you aren't ready and if you do it anyway, you will only regret it, and hate yourself, and despise me, and throw all this information in the garbage can. Don't do that. Take it at your own rate. Your god will naturally evolve you, as the alchemist, in the direction that you need to go, and what needs to fall away from you will fall away. And you won't have to start making things happen, things are just going to happen. You don't have to make it happen when you engage Consciousness and Energy, you are making it happen in the process. You have to make choices; that's your part. That's the role you play, and those are your doors that you have to walk into.

There are choices to be made after beginning this work. Remember, the people in your life don't own you. They don't own you. They're not the lord of your being, and they're not dictators of your destiny, and they're not going to die for you. You are alone in these decisions. And the more you go forward, the greater the love will come to you, to give you understanding. It's called grace.

And if people pass out of your realm, allow them to. By grace, you will have the strength to love them unconditionally. Let them go. And if they don't like what you're doing, have the grace to understand they don't understand what you're doing. They just lost a mirror to them; that's what they lost. Because you know, the people in your life never really see you. They see only themselves in you. You've slid out of their reality and they want you back, so they can communicate with you. If you go back, you'll step back into the image, because the image is the mirror to their reality. And you have to make choices on that issue, because above all, you have to be true to yourself.

Regardless of anyone and anything else, transfiguration is truth to self. And you have to honor that, no matter whether anyone else honors it or not. You have to.

"In my Father's House there are many mansions, and I have gone to prepare one for you." There is truth in that. In consciousness, and beyond these stars, there are many mansions and many adventures. There are many things that you could not begin to conceive. There is life of the most diverse forms that you would have to experience to know. And mansions of joy means houses of consciousness. You leave behind all the suffering and pain. It's true that in heaven there's no pain. And no suffering. And no hurt. And no disease. Because heaven is the consciousness you flow to when you are transformed and you leave all of the illusions behind. As it rolls up like a scroll in the last days, as a finished part of evolution, then you leave no tracks.

So you're not moving toward something that is so nebulous that the risk factor is so great as to make you run back to the pain of the past. You are going to an unknown, to a consciousness where adventures beyond this flesh await, and that is the reward of a transfigured person.

Bottom-lining this, what you've been looking for all of your lives and what some mediocre teachers have endeavored to teach, is the resurrection of the spirit. The resurrection of the spirit means you must be reborn and what that simply means is that you must burn away the image and create a new reality. You create a new kingdom. That's what it means. And in doing so, you are breaking down the image, the flesh, that it can be resurrected in consciousness. This is how you do it. You don't have to die on any cross. You have to be transformed. You've already died 10 million deaths before. You have to be transformed.

And every day, *be* the master. Before you eat, engage the consciousness, releasing energy and power. And there will come one morning when you look into your cup and a river of blue rhapsody will flow there, with stars twinkling bright that say "Come home."

You have a lot of adventures waiting for you. There are a lot of things you are magnetizing; you deserve them, because you created them. You will have to do a lot of creating, all on your own. You don't ever march in my army and not be equal to me. I can drive you, push you, but I know you're my equal. And one day you'll know that, too. And you have to trust, because I'm coming from where you're going. I know what the path looks like. I know what the view is about. So if I seem ruthless, it is personal to you. It's all for that glorious moment when transfiguration of the human body is God awakened as Self. And all that is called eternity upon eternity is created in that moment in consciousness, for you.

Do aliens exist? Yes, they do. And do aliens exist here in your world? Yes. I understand that when you cross from one country to another, you are considered an alien.

Do these beautiful ships exist? Yes, they do.

Have they been connected and intertwined in your history? In the fabric and tapestry of civilization? They are in its essence.

Depending upon which ones were making the assault, it has always been for the glory of the human being, for God.

First and foremost, you should know what you are. You should know that before you ever start looking around to discover anyone else. Because first and foremost, you are made up of a cosmology that defies planetary explanation, because it is the essence of planetary explanation.

It is called God. It is consciousness and energy. The life force.

Your intelligence and your capacity for intelligence is equal to anything that lies in any star system, in any dimension. You have the keys to be it all.

Knowing yourself first, before you go around looking under rocks and bushes and brambles is very, very important.

These entities exist. Their greatness is legend. What they are capable of doing is awesome. And what they can share with you is a word called magnificent.

But there is a law that intertwines all reality. Reality is relative to its creator. Whatever you are in consciousness, you equalize to you in your life. It is what constitutes your reality. And that law is synonymous with everyone. No matter how they look. No matter what they ride. No matter where they live.

The aliens we've been discussing understand it perhaps better than anyone on this plane. Because they have learned to captivate those laws and use them technically.

You are learning the law of magnetism. Equalization. You are only going to have an experience equal to your own greatness. You are only going to have an experience equal to what is relative to who you are.

And if your consciousness bears fruit, by continuously becoming aware of self and expanding, you equalize to you the same, no matter what dimension, no matter what time, or reality. You just equalize it.

A great person's life is dusted with the unknown. Because what makes them great is their consciousness and that is the result of their work, their intelligence, and the quality of the life they live.

The eccentric is more highly evolved than the civilized.

You're not crazy to want to engage a ship of light. You're daring! You're not neurotic to want to see wee people. You're bold! And you're not insane to take a ride on the light. You're fortunate.

Consciousness and energy are the keys that unlock every door of reality. The unknown is the greatest adventure awaiting you and that unknown truly, truly transcends these ships and their occupants. The unknown is precisely where these ships and their occupants are headed. The unknown is as close to you as it is to them. It is a moment away in realization.

The unknown is, I assure you, not boring. It's alright to want to be a part of these things and learn about them. (Remember, victims always draw victimization dependant upon how much energy they give to being the victim.

That is what determines how great their tyrant is!) When you engage an experience, it will be exactly equal to what you created in consciousness.

You deserve the right to understand these entities, these aliens. Because when you learn to love, it truly does transcend everything. That wild, enchanting, captivating, unrelenting, possessing, irrespective essence called love can even bring aliens together. And its seed, in consciousness, is its life blood.

Learning to love yourself and growing up as gods entitles you to the right to love others, to engage the whole universe of beings very similar to you. And the only thing that ever separated you was your ignorance and your intention of ignorance.

It was never important what you did. It was what you were and what you are. And no one will ever love you like you will love you through all the veils.

You are going to build bridges from this book. Wonderful, exciting, adventurous bridges, because you are creating a magnetism that doesn't just reach over the horizon, but beyond the sun and the North Star.

Wonderful things happen to you when your energy shifts.

Energy shifts don't happen because you have a "past," because you carry around your history. They happen when you transcend the past and become a fully participating, wholly activated now being, when you give off that kind of energy that comes from a now place, an eternal place. And as long as that energy is flowing from you, you are in eternity, and wonderful things happen as a result of it.

Illnesses go away. Joy comes into your life. Flowers start to bloom. Bees make honey.

When you have fully embraced this material, your life is going to change and become very filled. You won't continue to dig around under every rock, looking for truth. Because you will understand that truth is within you. And you won't search for happiness outside yourself. You will know it is within you.

You women will no longer dally after someone to take care of you. You will own the understanding that true security is within you.

And you men will no longer have to see yourselves in the role of provider. You will no longer need to show your strength to reveal your leadership for everyone to see in order to stroke your altered ego.

You will, instead, at long last be able to say that God is within you. And that is the greatest ego there is.

And what is the legacy of this journey within?

Joy, I am happy to say, at last, is your legacy.

So be it.

Afterword

Other Books Based on Ramtha Material

Ramtha
An excellent compendium understanding covering the cornerstones of Ramtha's teachings. Impeccably edited by Steven Lee Weinberg. *$19.95 hardcover.*

I Am Ramtha
The first high-quality, full-color photographic book on a channeled entity and his teachings. *$9.95 hardcover.*

A State of Mind, My Story: Ramtha, the Adventure Begins
JZ Knight's moving autobiography in which she tells the story of her life and her involvement with Ramtha. *$5.95 hardcover.*

Becoming, A Master's Manual
Edited by Khit Harding, this book provides a collection of quotes from Ramtha's teachings. Beautifully designed and presented. *$14.95 quality paperback.*

Manifesting, A Master's Manual
A collection of Ramtha quotes, beautifully edited by Knit Harding. *$7.95 quality paperback.*

Destination Freedom: A Time-Travel Adventure
Douglas James Mahr's inspiring adventure exploring self-discovery, under Ramtha's guidance. Stage I and II. *$5.00 each quality paperback.*

Voyage To The New World
A collection of Ramtha's teachings and commentaries written by Douglas James Mahr. *$4.00 quality paperback.*

Ramtha Intensive: Soulmates
Transcribed from Ramtha's January 1986 audience on soulmates, love and relationships. *$10 quality paperback.*

Ramtha Intensive: Change, The Days to Come
Transcribed from Ramtha's May 1986 audience on man, the environment and what nature has in store. *$10 quality paperback.*

Last Waltz of the Tyrants
This book is critical to understanding where we are environmentally, politically, and economically. It is a cry for change, a paean to sovereignty. Edited by Judi Pope Koteen. *$7.95 quality paperback.*

Financial Freedom
This book gives you the information necessary to implement the knowledge gained from *Last Waltz of the Tyrants*. If you want to be financially sovereign, this

book will show you how to make that happen. Edited by Judi Pope Koteen *$9.95 quality paperback.*

Additional Reading List

Above Top Secret	Timothy Good
Moongate: Suppressed Findings of the U. S. Space Program	William L. Brian II
Dimensions — A Casebook of Alien Contact	Jacques Vallee
The Twelfth Planet	Zecharia Sitchin
The Stairway to Heaven	Zecharia Sitchin
The Wars of Gods and Men	Zecharia Sitchin
The Lost Realms	Zecharia Sitchin
Genesis Revisited	Zecharia Sitchin
Occult Conspiracy	Michael Howard
None Dare Call It Conspiracy	Gary Allen
UFO Contact From Andromeda	Wendelle C. Stevens
Alien Magic	William Hamilton
UFO Crash at Aztec	William S. Steinman and Wendelle Stevens
The Unseen Hand	Ralph Epperson
An Alien Harvest	Linda Howe

How to Order Books

The above titles represent a selection of books recommended for further study. There are many, many important books. This list represents books available through RSE Products and Services, a division of Ramtha Dialogues. If these titles are unavailable locally, you may order them through RSE. All Ramtha materials and hundreds of other books, videos, and audio tapes are available as well. They accept no credit card purchases. You may reach them 9-4 Pacific time, Monday through Friday. They are closed during weekday Intensives and events. Due to market fluctuation, contact RSE for current book prices and shipping rates.

RSE
404 First Avenue S.E.
P.O. Box 519
Yelm, WA 98597
Mail Order Phone Number: (206)458-4771

Ramtha titles are available also through:
Beyond Words Publishing, Inc.
Box 492-B
Hillsboro, OR 97123
1-800-284-5109

To experience Ramtha in person, or on audio and video tapes, contact:
Ramtha Dialogues
P.O. Box 1210
Yelm, WA 98597
(206) 458-5201

For additional copies of *UFOs and the Nature of Reality*, contact: **Indelible Ink**, P.O. Box 1415, Eastsound, WA 98245. Or call 206-376-5184.

UFOs and the Nature of Reality is available as a Book on Tape. Narrated by Judi Pope Koteen. Original music by Daniel Ferris. Contact Indelible Ink for ordering instructions.

Publisher's Note

This book is living proof that dreams can come true. We hope it is only a beginning. Our plans include a periodical and additional titles. If you are interested in information on upcoming publications, or in sharing your personal story, please contact us through the editorial address below.

As Ramtha says, we are all threads on the tapestry of life. Perhaps by sharing our stories, we can see the overall canvas more clearly.

Please order additional copies of *UFOs and the Nature of Reality* directly from us or encourage your local bookstore to keep several copies on hand. Thank you.

Indelible Ink
Business Address
P.O. Box 1415
Eastsound, WA 98245
206 376-5184

Editorial
P.O. Box 33
Orcas, WA 98280